Of LOBSTERING *and* LOVE

Trials and Triumphs

Roy P. Fairfield

Bastille Books
Saco, Maine

Bastille Books™
P.O. Box 1648
Saco, Maine

Library of Congress Catalogue Card Number: 90-82742

ISBN: 0-9621921-2-0

Cover Design by Dan Howard
Design Sense
Portland, Maine

TO:

    Maine Coast lobstermen and women, both past
       and present who have risked
       their lives and fortunes
        but especially

    to the Rumery Family who have shared
       their experiences and insights
       so generously...

Table of Contents

Chapter 1

Thund'ring Reality

Now Harold was alone with his father's boat, grandfather and dad buried a week apart. The old house, the mangled wreckage of his Gramps' vessel and the long windrows of lobster traps, grey-lustered in the early May sun, now his.

As he backed his Ford pick-up into the main road, the enormity of the load he now shouldered struck him. While Hal had sometimes mused about whether or not to carry on the family tradition without the older men, little did he dream that catastrophe would strike so quickly, both killed by the same storm that spawned the rogue wave, though the younger man valiantly fought a week for his life at the local hospital after his grandfather drowned. Nor could he fathom the fates that sent him to the dentist the day they capsized rather than with them to Jocks Rocks and the outer regions of the Baysville Harbor lobstering grounds. As he shifted his truck into second, Hal reviewed his feelings of grief that he experienced when both his mother and grandmother had died of cancer just months apart two years before...also the anger when his only aunt, who lived in California, had told him at her father's funeral that she had no intention of filing any claims against the will-less estate so long as Hal continued the family fishing tradition. He had no particular love for his Aunt Sarah; yet didn't want to fight her even in his irritation that she was holding him hostage to their family heritage. He knew that his family had a reputation of keeping oral promises, one that every generation boasted about, back to their New England origins. He knew, too, that he was supposed to live up to the reputation of his family's folk heroes, some of whom seemed larger and stronger than Paul Bunyan and Ahab of Moby Dick fame.

Much as he disliked lobstering and feared the power of the sea, he felt compelled by tradition and "his word" to continue the life he had begun as a child with his weather-hardened father and grandfather, first on both their boats, then mainly helping Gramps when arthritis made it difficult for him to manage the winch and warps, especially during the hard Maine winters. He knew, too, that though he was considered "different" than most of the boys with whom he'd grown up in this small fishing village and that many in the powerful lobster gang didn't like him, still the sea was mixed with his blood. I *"know* it," he repeatedly observed, "but I don't *feel* it."

He had followed his parents' wishes, commuted by bus and car to Statesville for four years, finished high school routinely at eighteen. But he facetiously remarked in a letter to his Portland cousin, Sue Curtis, "I went to school with books and sketch pads in one hand and lobster bait in the other."

Both parents and most residents of the tiny fishing village knew his heart was not in it, and they resented his attitude and impractical interests and skills. Teachers constantly caught him drifting away from classroom topics, writing haiku and free verse and illustrating them with sketches of flowers, waves, dunes, driftwood, fish and other seaside objects. Too, they frequently found him daydreaming as he sketched boats of every size and description, doing it behind his assigned texts and papers; they forced him to hand them his five-master clipper ship, sketches of every part of a lobster boat, and his streamlined dories. In fact, he thought he'd probably designed more boats in his classes than had ever been built on the Maine Coast. A few teachers moved him into the front row where they tried to intimidate him to cease daydreaming and drawing. But he sketched as others might doodle, sort of "in a zen trance," as he explained it; he heard classmates reciting or causing a commotion even as he drew. But when he saw a teacher coming toward him, saying nothing but holding out a hand, he knew he'd been caught. When, at the end of his sophomore year he asked to have his sketches back, the teacher remarked, albeit sarcastically, "Harold, we should present them to the town library and hold an exhibition!" He smiled through confusion as he held out his hand to retrieve the fat portfolio, because he wasn't quite sure what the teacher was telling him. Many times, he thought, "Perhaps I should tell them that I'm going away to college to become a marine architect or artist? Then they might encourage me?" But nobody even suggested it. Rather, they treated him as a classroom stumbling block. Nor did his classmates give him much support. All but three of seventy-two voted him "Most Antisocial and Least Likely to Succeed in Chosen Occupation." Only one young woman, Maria LaRoux, sought him out as he talked with his mother and father after the graduation ceremonies; she shook his hand and wished him well, remarking, "Hal, I don't agree with our classmates," then rushed quickly away to find her parents.

But once on the road with his pick-up, headed for the docks on that cool May morning , he vividly recalled Maria's soft round face and jet black, boyish-bobbed hair as he shook his head vigorously to disentangle the cobwebs, ran his rough hand over his blond stubbly beard, and tried to enjoy the beauty of the day. It was sunny but snappy. Lawns were still covered with frost, and he noted one boulder whose lea was still dappled with snow from the previous week's storm. NOAA weather predicted a fair day, 15-knot SW breezes by noon, so he knew that he had little time to lose if he were going to tend his "inside" traps, reset only two days before, five days after the funerals. As he passed his friend Warren Reed's home, he was surprised to see his Chevvy truck still parked in the yard. "Strange," he thought. Reed was notorious for being on the water at dawn and back at the docks by late morning. "What's with him today?" Hal thought, craning his neck to peer behind Reed's back shed. But nobody was in sight. "Strange," he mused, again.

At the general store, Randy Ballou flagged him down.

"Hey, Hal," Randy blurted, "Got any extrah bait?"

Hal knew that Randy didn't much care for him, regarded him as a poor fisherman, been quoted as saying, "another one of those damn Yankee families." There was also a ten-year age difference, and he'd overheard Randy call him a "damned stuck-up arty kid." But Hal was polite, answering Randy with a question, "Shuah, how much ya need?"

"Oh, 'bout a barrel or two, can ya spare it?"

"Shuah, hop in, and I'll fix ya up."

So Randy got into the pick-up with a quick "Obliged."

Hal opened the conversation, "Say, Randy, what's with Reed? Saw his truck in the yard!"

"Not shuah, somebody said his son-in-law, Andy, was in an accident two nights ago. Maybe he's at the hospital?"

"Isn't he the guy who's into computahs at that big plant down 'n Lewiston?"

"Ayuh, I think so."

The two men continued to the docks, riding in silence. Hal felt that Randy would have liked to talk about his dead father and grandfather, but said nothing...until they were leaving the truck.

"Look, Hal," Randy yelled above the roar of boats taking off for lobster grounds, "Sorry 'bout your dad and gramp. They were good fishermen. Hated to see them go. You gonna keep fishin' in their shoes?"

"Thanks, Randy," tears welling into his eyes. "Yup, I'll follow the fam'ly tradition as long as I can stand it. But you've heard me say befoah, I don't really like it."

"Yeah, I understand. There are days when I wish'd I was doin' sumpthin' else. It's a tough life."

"Ayuh," Hal observed, "and all these damned tourists come up heah from Massachusetts and New York and think it's romantic to see all these pretty boats with their little triangular sails.   And we're the poooah bastards who go out and risk our necks..." At that, tears welled up in his eyes again.

By that time the two men were walking toward the baithouse together. Hal took out his bandana, blew his nose, wiped his eyes, never looking up to Randy, but saying, "Sorry..."

"It's O.K., Hal, yer right, we risk our necks, get lousy prices for our lobsters, and our marketmen make the dough. Don't I hate these stoahmen who get fat off our asses! Maybe you'd be bettah off goin' inta commercial art?" He gave Hal a sidelong glance to get his reaction.

Hal said nothing, but as they approached the baithouse, Hal began laughing. "Gads how that bait stinks this mornin'. I wonder we don't all turn into rotten fish." Randy joined in the laughter as they rolled out two heavy steel barrels, with Hal's father's initials, "DD", in huge letters on their black sides.

"Look, Hal," Randy remarked, "my skiff's right heah. Let me row ya out to ya boat and it'll save ya time this mornin'."

"O.K., can use the time, Randy, NOAA says breezes'll be pretty stiff by noon. I'm not goin' much beyond the fifty traps I set on Tuesday just past Jocks Rocks. Dad and Gramp always fished close to that area, so I thought I'd begin there. And you? How far ya goin'?"

"Three, foah miles off toward Poahtland, where I usually fish this time a year. Gotta get out 'n' back befoah she starts a-blowin'!"

The two men got into Randy's skiff and moved smoothly out to their anchorage, neither saying much but looking around to observe the nuances of the harbor. Both knew the land and harborscape so well they could detect the slightest overnight changes.

"Looks like Earle's takin' watah, keeled over a bit," Randy observed, as they passed a boat whose stern carried the boat's name, "Artie K," in large black letters.

"Told me he hit a reef the other day, right after the big storm, split a plank, but he hates to take her out for replankin', just barely makin' ends meet."

"Ain't we all, Randy?"

"Suah enuf!"

Reaching their boats, the two men started their motors with a roar, went directly to the dock to hoist their bait aboard, then headed for open ocean. Hal cast off first.

Standing at his steering wheel, Hal measured nearly six feet, was both brown and burned from days facing sun, rain, sleet and wind. Although he weighed only 165 pounds, he was muscular and could leverage lobster traps as though they were children's toys. His sandy brown hair and dark brown eyes softened the premature wrinkles just below his temples, lines drawn by constant squinting against the sunlight. Hardened by the weather, he tended to go bare-armed most months of the year and loved his denim shirts for the protection they gave him for rubbing against the traps' oak slats.

As he shifted the throttle into top speed, Hal felt a sense of well being despite all that had happened to him so recently. And much as he feared the sea, he knew he had the skills to be a successful lobsterman although he had a reputation in Baysville of being nothing but a leech and a wimp since he'd always stood in his family's shadow. Not only that, the previous winter he'd had two costly accidents, doing serious damage to three other boats

when maneuvering at the docks.   This only confirmed
what folk in the village "told around," that Hal preferred
day dreaming, drawing and poetry to the rough and
tumble of a fisherman's life. So they concluded that "his
dreaming caused the accidents" and Baysville would be
better off without him.   And despite their words of
sympathy over the double death that occurred so
suddenly, Hal felt the hostility "as thick as pea soup"
whenever he got to the docks.

Though he sometimes dreamed about other lives
he might live, he did enjoy the sense of adventure and
freedom of being on the water. "At least," he mused, "I
don't have some bull's-ass teacher or foreman telling me
what to do. It's now my boat and my territory out
here...even if I barely make enough to live on. I'll fix up
the old family house for my needs, collect the insurance
on Gramps' ruined boat, split the money between a CD
and a money market fund and have a little to fall back on
until I get squared away. Maybe I'll build a couple
hundred more traps, add a sternman and see if I can do
better than break even."

His reverie was broken by a loud horn to his
starboard. He glanced to the right under his long black
visor to see Randy, at max speed, passing him with a
thumbs-up gesture. He waved back, holding up his
fingers in a V, a ritualistic signal Baysville Harbor men
used to wish others, "Good fishing!" If one didn't get that
sign, he knew he was on the other's Shit List. As he
watched the stern of Randy's boat disappear in a feathery
fountain of foam, Hal wondered when in May Randy
would bring his traps closer to shore. Most fishermen
regarded him as one of Baysville Harbor's most successful
lobstermen, no matter how much poverty he pled. It was
something like "monkey see, monkey do" for the men
who did not have quite the feel for the bottom despite

their high-falutin' SONAR equipment.   Hal also wondered about that equipment, now that his dad and grandfather were gone and could no longer resist his own feeling that they should modernize. He guessed that it was about time for Randy to reset his traps somewhat closer to shore than he had fished all winter and spring. He also knew that he was taking some risk in using Jocks Rocks as his current location since another severe storm might wipe him out. He didn't have an exact trap count since his family had taken that to their graves, but he estimated that they'd probably lost at least seventy-five traps around Jocks Rocks during the recent storm. Scraps of paper they'd left in the old workshed where they made their traps suggested that they might once have had as many as 700 altogether. He'd long since given up asking them how many they were fishing.   While other lobstermen might boast of their total, especially if over 300, both his father and Gramps kept their count close to their long yellow slicker-aprons. And his own effort to count them as they fished was frustrated by their methods of random setting. Whereas many of their fellow fishermen set traps in lines or even on strings of three, five or ten, both of his seniors preferred to set one trap at a time and in "some of the damndest holes" he could imagine. He always had felt that they fished too close to shore, took enormous chances, especially in one crevice they called "Dinah's Crack," where legend claimed that a black woman had drowned a hundred years before. In fact, he guessed they lost their lives because of these habits which both said they'd learned from their own grandfather who had learned it from his great grandfather. He suspected, too, that his own reputation of being a wimp and "stuck up" related to his having openly shared his Yankee family history to a couple of peers he thought he could trust. Naturally, they blabbed!

Soon reaching his strings of traps, he tried to stop all musing and concentrated on the routine of locating the leaders or toggles, pulling in his black and white striped buoys, casually checking the knots and the integrity of his father's Licence Number, 77-DD on each buoy, then inserting the warps into the open pulley above the hydraulic winch, and powering the trap to the surface. In many ways, he thought, "it's all so automatic when things are going right." Yet, despite his having only recently reached twenty-one, he knew from experience that anything could happen to make life difficult if not impossible. Ropes got tangled in the prop; marauding and jealous neighbors for other gangs cut lines; the boat had "a zillion things" that might go wrong and often did, despite systematic efforts to maintain it. And the morning might be bright, with sheets of sun splinters sparkling on the smoothly undulating waves, but the ocean's power had to be respected and the weather's dictates obeyed. As he tended his traps that morning, he recalled some of the superstitions Baysville fishermen turned into rituals. One fellow never went onto his boat from the port side or watched Canadian geese fly to the horizon. Another insisted on using a bowline knot and no other, from one end of his operation to another. "Half-hitches are for sissies!" he exclaimed. Hal knew, too, that some Baysville men refused to use their ancestors' buoy colors, one noting, "Ain't tradin' on reputations of the dead!" Hal mulled over the propriety of using his parent's markings. Baysville Harbor was a bubbling cauldron of "war stories" and legendary figures who became folk heroes for challenging both sea and tradition.

"But, of course," Hal muttered to himself, "The most discouraging thing about this whole business is opening a series of empty traps." And that's what happened on that particular May morning. One after another, empty. By the time he paused for his mid-morning peanut butter sandwich and Coke, he'd pulled twenty-three traps and had eight "keepers." "May as well go home," he speculated, knowing that he had not yet earned enough to pay for his gasoline. He decided, however, to continue. He feasted on the ocean colors and the long low hills behind The Harbor and the way the town's one white church spire rose from the evergreen forests along the shore.

Hal also felt a certain comfort and safety by the presence of his lobstermen acquaintances fishing a mile or so outside of him. While he felt their hostility for all the reasons he'd repeatedly turned over in his mind, he knew he could turn on his CB and call for help in a n emergency...or if push came to shove in confronting another gang. Normally, though, he preferred to fish without the clatter and jiving of the radio. When all was quiet, he could dream, draw sketches in his mind, write an occasional haiku on a new trap and push it overboard and watch his pencillings sink through the distorting water, consider his future, wonder about the nature of life and do routine chores more or less automatically. When he found himself "slipping out of time" in his favorite zen sense, he would jolt himself back to the boat for what he called "reality checks."

By midmorning that day he noted that the wind had begun to spring up, saw Baysville boats as well as one from neighboring Barnsville, down the coast, unwrap their tiny triangular sails to hold their boats into the southwest wind. But he decided to fish awhile longer without the sail, enjoying the way the tides swirled around his boat, the "Mary Q," named for his mother. He also thought he saw and felt more of the ocean and shore when his boat rolled in the winds and tides. Waves dashing on Jocks Rocks also brought back many memories. His decision about the sails reminded him of discussions with his grandfather who claimed he didn't need such sails to be a good lobsterman. During his many trips with Gramps, he'd "learned the ropes," heard the history of his family, how they'd come to New England in the seventeenth century, survived famine and Indian captivities, followed the sea both as fishermen and sea captains, traveled around the world, harpooned whales on the seven seas, married their "own kind," been cut badly by lobsters and dogfish, suffered death and life and life and death in all their bewildering varieties. Hal also valued the many pieces of wisdom he'd learned from the old man but often wondered how he could use them in a world seemingly shattered every day by some world event such as Three Mile Island, Iran Contra, Fall of the Berlin Wall or the American government's seeming disregard of our nation's astronomical debt. In fact, he found himself frequently using such Gramps expressions as "homely as a stump fence," "the man will live as long as a crowbar," and "the stuffin's the best part of the fish." But Hal felt no need to investigate whether or not such expressions were original with Gramps, had come down through the family or were part of his Maine heritage. "What difference does it make?" Hal mused. "Gramps enjoyed them. I like 'em. Isn't that enough?   even if the U.S. goes bankrupt tomorrow?"

He also reviewed dialogues he'd had with himself the previous week, "to stay or not to stay? What other occupations to pursue?" But he always came to the same conclusion: "Dad and Gramps wouldn't like it if I left." Too, he remembered the vehemence with which Aunt Sarah intimidated him at the cemetery (even in the presence of the two caskets) screaming about his "duty to the family heritage." As he turned off the motor to drift a bit while eating his sandwich, he suddenly stood up and screamed to the wind. "Goddamn it, why did I promise Aunt Sarah anything? Why in hell should she control my life from far-away California? What would she get even if she put some shyster lawyer on the trail of the two estates? And why did I let go of my own life?" The anger, the smell of peanut butter and lobster bait stuck in Hal's throat, in fact, caused him to throw up. This made him the madder even as he asked himself, "Can this be the first time I've ever been seasick?"

Returning to the helm, however, and restarting the boat, he took a couple of swallows of water and several quick breaths of fresh air and got back to fishing. Fortunately, traps east of Jocks Rocks yielded more "keepers" than he'd gotten all morning so this momentarily took Hal's mind off his dilemma, what he'd called "being in prison." Finally, when the wind stiffened, he put up his sail, finished tending all the traps he could reach at low tide, then re-wrapped the sail and revved his engine back to harbor with forty-seven lobsters which he estimated would pay for his gasoline and bait and give him about fifty dollars to boot.

Back in the harbor he transferred the day's catch to his cage under the boat, hopped into his skiff and rowed ashore, hoping none of his fellow fishermen had seen him throwing up. Taking off in his truck, he made a quick stop at the store to get milk and a sub sandwich, then drove directly home.

Chapter 2

Hard Decisions

The next week was a difficult one. He hit a ledge and bent the boat's drive shaft forcing him to ground it in the wharf repair area on a very low midnight tide. Since he could find nobody to help him, he did it by himself, floating an old family cradle under the boat at high tide, letting the boat settle on the beach when the water went out. The next morning, a Saturday, as he worked under the boat removing the shaft, a gaggle of squawking boys came along to taunt him, chanting "Hal, Hal...'s got no pals! Hal, Hal..."

This infuriated him, but he listened for awhile, hoping they'd get tired and leave. But they didn't so he crawled out from under the boat and screamed, "Go home, you guys," then regretted it, worrying that the kids would "tell stories" that their parents would believe and come back to threaten him the more. Sure enough, Bill Watts, father of one of the boys, showed up.   A lobsterman a few years Hal's senior, he pointed a finger and shouted, "Hal, don't you ever lay a hand on my kid again!"

Very quietly, without even looking up from his repair job, Hal responded, "Bill, I didn't touch your kid."

"He said you did."

"Then he's lying, Bill."

"That's a heavy rap, Hal."

"Sure, so's his. He was part of a gang taunting me and I told 'em to go home. I don't know why they're picking on me unless it's because you and some of your buddies want to drive me from Jocks Rocks."

"Nonsense, Buddy. But the guys don't like you. Yer a no-good lobsterman and a menace in the harbor. You always traded off your old man and grandfather, didn't want to fish anyway. Dreamed your way through school. Why don't you get smart and go back to all you know, poetry and art... before the boys rough you up?"

Hal said nothing, continuing to dismantle his prop cage.

Finally, Bill called Hal a "fuckin' coward" and stomped off. Hal continued with his repair job. But his eyes filled with tears as he worked, feeling both sorry for himself and angry with the very gang his family had organized. "Surely," he mused, "I didn't ask dad and Gramps to die!"

Fortunately, he removed the shaft with no further interruptions, but he fully anticipated more harassment. He knew that Bill "told it like it was." Staying in Baysville Harbor might be dangerous. As he left his boat he made sure that all lines were secured and his cabin locked. Loading the bent shaft into the back of his pick-up, he headed for Statesville where his father's friend, Lin Dayton, ran a machine shop.

As he entered the shop, shaft on his shoulder, Lin greeted him cheerily, "Whatsamatter, Hal?"

"Bent shaft, Lin. Hit ledge at Jocks Rocks."

"Dang'rous place to fish, Hal. Maybe worth the risk though, happened to your Gramps lots of times." Lin picked up the shaft with one hand, laid it on his metal bench, rolled it back and forth over a steel plate to check the bend, concluded, "Think we straightened it before."

"B'lieve so."

"Hal, you'd better get a new one, it's crystallized, pretty well gone.   But it'll take three or four days to get it shipped out of Portland...unless you want to drive down and get a new shaft on Monday morning. I could machine it early afternoon."

"O.K., give me the exact dimensions and I'll fetch it myself. Can't afford to keep the "Mary Q" outa water too long, gotta keep fishin'." He didn't tell Lin that he was worried about the safety of his boat.

"Sorry 'bout your dad, Hal. Good man and great fishermen. Tough to lose both of 'em at the same time."

"Thanks, Lin," Hal responded, blowing his nose.

"Sorry," Lin comforted, touching Hal's sleeve.

"I'm O.K."

"Rumor has it, Hal, you may give up fishing."

Hal wondered who'd been talking, but observed, "For lots of reason, I can't...unless the gang drives me out. They'd like nothing more. That would leave Jocks Rocks for them 'cause they don't believe my family had prior rights there."

"Goes back a couple of centuries, doesn't it, Hal?"

"Yup!"

"Then hang in there. I know that gang. I occasionally see them here when they need machine work done. 'Tween us, some of 'em are pretty tough. If they know you're afraid of 'em, they'll bite you like a dog."

"Wish I didn't know that so well, Lin. They've begun."

As Hal got into the truck to head for home, Lin shook hands firmly, remarking, "Keep up your courage, Hal. Find a way to make them fear you. See you Monday afternoon."

Sunday was usually a quiet day for Hal. For years he'd followed his parents' example, rarely working on anything but household chores, made sketches or sharpened tools in the back shed. So he slept late, read the Maine Sunday TELEGRAM delivered in the white plastic tube across the road from his driveway and quietly eased into the day, planning only to "touch up" a circular saw for ripping up some oak planks for new traps he might build.

About eleven o'clock the phone rang. It was Randy Ballou. "Hi, Hal, seen your boat this mornin'?"

"No, sumpthin' wrong?"

"You gotta see it to believe it. I'd meet you there if I weren't 'tween a rock and a hard place. I just can't be seen with you, gang wouldn't like it. But you've helped me a couple of times, so this is a friendly call." He hung up abruptly.

Juices began stirring in Hal's stomach as h e snatched the truck keys from the kitchen counter and rushed for the boat. As he drove the mile between marsh and beach, he kept repeating, "What 'n'ell happened? Damned gang's gonna get rough!"

As he turned the corner into the wharf, he saw the damage. Using a can of yellow paint, somebody had sprayed graffiti over the entire bottom of his vessel. Walking from his truck to the boat, he could read:

COWARD....DAMNYANKEE...NO GOOD FUCKIN
FISHERMAN...WHY DON'T YOU LEAVE TOWN?

Included, too, a skull and crossbones. Looking on the scene: two local boys, about twelve and an elderly man-woman couple he figured must be tourists, confirmed by their New York license plates. The woman spoke up, "What's this about?"

Hal looked sidelong at the two boys, knowing them as tattlers, and calmly replied, "Don't know," even as he felt "the urge to kill" those who did it.

The man from New York continued the query. "Are you going to report it to the authorities?"

Hal quickly retorted, "Prob'bly they know it already. News travels fast here on the Coast. In fact, by now they prob'bly know it over in Barnsville!"

Hal left the couple staring, mouths open, as he walked around the boat to survey the total damage. He knew that he dared not report the matter to the local deputies. Rumor had it that they were probably "bought" by the leaders of the local lobster gang. He knew just enough about Baysville politics to realize that many of these men hated his parent and grandparent because of their two-century heritage and prestige, their power as Baysville "strongmen," also skills as lobstermen. This had been a topic of many dinner conversations. His mother trembled as she warned his father not to do anything violent, praying for "peace" while saying grace. Suddenly, all of those anxious childhood moments came rushing back, and he wondered if his artistic talent and passive nature came from her?

While surveying the boat, he was aware, too, that his chief nemesis, Bill Watts, who lived in a ramshackle, weather-beaten house overlooking the wharf, was standing in his window watching for Hal's reaction. Hal felt the man's eyes burning into his flesh, hotter than the noonday May sun. So Hal casually rubbed his hand along the bottom of the boat, pleased that the graffiti artist, whoever it was, had restricted spraying to the red bottom and not messed up the "Mary Q's" white sides. As he "went through the motions" of examining the boat and considered what to do about it, he stood on tiptoes to check the cabin and saw that nobody had been aboard. As he contemplated action, a small knot of local men gathered at the port side.

There was a considerable amount of joshing among the group, but only one, a high school classmate, Sal Lugano, spoke directly to him, "What do you make of all of this, Hal?"

Hal was tempted to say nothing, for Sal was a ring-leader of his class's vote predicting his future. Also once during the previous year Sal taunted him as "Gramps' Boy." But Hal knew, too, that it was better to respond than remain silent, so he said, obliquely, "Somebody's heard about my dad's and grandfather's deaths." And he started back toward his truck.

This confused Sal, but he shouted, "Aren't you going to do something about it, you wimp!"

Hal stopped at the edge of the dock, turned part-way around, and said matter-of-factly, "What can I do? They're already buried."

Back at the house Hal again debated what action to take. "Sell out, leave and do something I really want to do? Fight 'the bastards' to prove to them (and me) that I'm not a coward? move my field of operation to Barnsville? After all, it would only be a nine-mile commute and I know there are lots of guys there who respected and loved my elders; nor does my family history threaten them." Hal also wondered if he should sleep aboard his boat that night, use his Polaroid to catch the next wave of despoilers red handed? He felt certain that it was better to leave the graffiti on the boat than try to paint it over; he was fairly certain that that would invite further nuisance. Hal even asked what he called "the cosmic question," "Why am I, Hal Daniels, the person I am here in this neck of the universe?" and "Should I move far, far away, Aunt Sarah be damned?"

"If only I had someone to talk to," he mused, feeling the emptiness of his parent's loss the more. On an impulse he decided to go to Barnsville to see his father's old friend, Jerry Moulton, also phone his Cousin Sue and try to see her in Portland the next day. He phoned Sue, got her answering machine and asked her to call back. As he sat at the phone and speculated some more on what to do, he found himself sketching outlines of dories and sand dunes on his scribbling pad. This drew him back into his center so he picked up the pad and took it to the porch where the sun was not only turning the space into summer but was also pouring into two trays of tomato seedlings he'd planted in April. Within moments he was "lost" in his sketches, and they simply flowed out as though magic. After staring out the window at a couple of diving gulls, he found himself writing a haiku as he did in school.

> Gull dipping toward dunes
> tips sharply into steep bank,
> mew mournfully,...Soul...

He could never understand why they flowed through his mind and pen like water, and for almost an hour he forgot his boat troubles and the dilemma of deciding who he was, what to do.

Because of this preoccupation he didn't see Elmer Kruse, the local deputy, drive into his yard.

When Elmer rapped heavily on the porch door, Hal got up to greet him, held out his hand, said, "Come in."

Elmer grumbled something but ignored Hal's hand. Hal invited him to sit down, but he wouldn't, snapping, "Gotta complaint from fam'lies at the dock, don't like the graffiti sprayed on your boat, want it removed right away, say the pornography hurts their children. What yer gonna do 'bout it?"

This angered Hal, but he remained calm, returned to his chair, leaned back, replied, "Why don't yer sit down Elmer, and let's talk about it."

Elmer stood fast.   "Nothin' ta talk about. Complaint's a complaint."

"But I think there IS somethin' ta talk about. You don't think I sprayed graffiti on my own boat, do ya?"

"Don't know. Cooooooouuuuld be?"

"Come on, Elmer, be reasonable. <u>You</u> know who sprayed the graffiti on my boat. Why don't yer ask <u>them</u> to cover it up?"

"No, Hal, <u>you</u> do it and <u>you</u> do it this afternoon. It's your boat, and I don't know who did it. Maybe it's your excuse to leave town since you know you can't fish worth shit!"

Hal heard the lie, took a long deep breath, stared out the window at a couple of gulls fighting over a crab in mid-air, watched them dive and swoop, heard them squawk. Regaining his composure, he asked, "And if I don't paint it, Elmer? What then? After all, I'll put the boat back in the water when I get a new shaft, early Tuesday morning at the latest."

"Look, I have a written complaint from four residents near the dock. You paint it today or I'll fine you heavy every day it's not done, under our town nuisance ordinance."

Hal was tempted to call his lie, but instead looked up at the man in disbelief, took another deep breath, exclaimed, "Damn it, Elmer, this is blaming the victim! Somebody messes around with my boat and I pay for it. I'm tempted to tell you and your rotten friends where they can go, but I won't. I'll paint the boat...this afternoon, but you'd bettah be damned sure it doesn't happen again! Now, please leave before I get real good and angry and do something I shouldn't."

"Are you threatening me?"

"No, you've done your duty. Politely, please leave."

Within a half hour Hal was at the dock with a gallon of red bottom paint, beginning a job he'd intended to do when he put the "Mary Q" overboard. Naturally, he drew a crowd of locals, including children of some of the toughest gang leaders. Again, they taunted him raucously. But the louder they became the cooler Hal remained. This made them the madder...until Bill Watts' youngest son threw a pebble at Hal. Luckily two things happened. The boy's aim was poor, and Elmer Kruse showed up just in time to see the rock fly.

He grabbed the boy by the right shoulder just as his father collared him from the left side. A pulling contest ensued, and Elmer shouted, "Bill, leggo or I'll arrest you."

"Elmer, you sonuvabitch, try it and I'll beat you up!"

The deputy was rattled but calmly replied, "I could take you in for that threat, but I won't. You filed a written complaint. Hal's painting his boat. Now take your kid and get to hell out of here." And he turned to face the entire crowd, speaking firmly, "O.K., you've had your fun.

Hal's doing what I told him to do. Go on home and leave him alone or I'll run you all in for disturbing the peace."

Several grumbled. Another boy picked up a rock, but his father knocked it out of his hand. One young mother snarled at Officer Kruse, "Ain't ya never heard of the First Amendment?" Kruse watched as they dispersed, then turned to Hal who said, "Thanks, Elmer, I was beginning to worry that they'd attack me physically...maybe chips off the old blocks?"

Kruse watched for awhile, then observed, "Hal, I don't think they'll bother you any more so I'm goin' home. Got some chores ta do for the old lady. But you'd better take heed. I wouldn't get any funny ideas about stayin' on board tonight. I'll drop around a couple of times late afternoon and ask the night officer to check your boat ev'ry hour or two to see if it's O.K. You damn well better get your boat in the water by Tuesday and reduce the risk of more mischief. What you do about the long run is your business, but don't say you ain't been warned!"

Hal thanked him and went back to painting, continuing to weigh all of the options before him while postponing his visit to see Jerry Moulton in Barnsville.

Next morning Hal drove off to Portland early, hoping to see Sue and buy the new shaft. At 8 he stopped at a roadside phone booth, figuring she hadn't gone to work. Sure enough, he caught her as she was about to leave, asked if he could take her to lunch. She was free so he began preparing for their dialogue. While they had never been close, she was his one relative living nearby. They hadn't fought since they were both eight, so he looked forward to a friendly chat.

His new shaft safely locked in the truck, he decided to walk around the Old Port, window shop, and casually amble toward Congress Square where Sue worked as secretary in a law office. The morning passed fairly quickly, and he was so busy looking in shop windows, buying art paper and browsing in book stores he didn't have much time to brood. Once meeting her and getting a warm hug, he quickly got down to his concerns in a small restaurant up Congress Street.

"I'm in trouble, Sue," he began, "but first tell me about you. How're you doing? How's your new job? We didn't get much time to talk at the funeral."

Sue could see that Hal was wrought up but responded to his questions as she looked into his deep-set brown eyes and concluded that he'd been missing sleep. In fact, she thought he looked twice his age, especially with his blond stubbly beard and weather-beaten face, but felt it best to say nothing about it, finally summarizing her own situation, "In short, Hal, I don't date much. My boss is wonderfully understanding about my slow grasp of legal terms, and I want to hear about you."

Hal paused, took a bite of his cheeseburger, then carefully outlined what was happening in Baysville Harbor, concluding, "Nothing too new, Sue, but this time it's happening to me and not dad or Gramps or friends.

Sue responded caringly. "I can empathize, Hal. I used to worry about Gramps, too, and knew that the scene could be violent...and I hate violence. But the question remains, 'What are _you_ going to do?'"

Hal outlined the choices he'd thought about, including both extremes, staying on to fight, "with guns if necessary," or giving up entirely "to seek a commercial artist's job...maybe here in Portland where I could see you more often."

Sue wrinkled her nose a little, said, "Hal, much as you may hate fishing and the sea, I think you'd feel like a fish out of water, if you'll forgive the pun, if you ever left Baysville. Competition here can be dog-eat-dog."

"Maybe so, but it might be worth a try?"

"But, Hal, didn't you promise mom you'd carry on the family tradition?

"Yes, but she twisted my arm to get the promise, and I still don't know all the reasons why. Guess it's mostly about loyalty to family heritage"

"I didn't know that!"

"Yup, said she'd challenge me for the estate if I didn't promise."

"What!"

"Yes!"

"But she doesn't need money, Hal. Husband's loaded!"

"I don't think it's money, Sue. Maybe power?"

Sue continued to munch her salad but said nothing. Finally, she looked out the window past Hal's ear, venturing, "I always knew that mom had a big ego and loved to boss people around, but I wouldn't have dreamed her capable of anything so cruel as to enslave you to an occupation she knew you hated."

"But, Sue, that's the way it is."

"Hal, give me a week to see if I can talk her out of it, give you more freedom of choice. Isn't that what this country's about, freedom of choice? In fact, I forgot to tell you I've joined the Pro-Choice abortion forces. Don't be surprised if you see me on television demonstrating in the streets or getting arrested,

Hal reached across the table and put his palm on the back of her wrist, exclaiming, "Good for you!"

Sue smiled, then asked again, "Shall I talk with mom?"

"No harm trying, I s'pose. Unless the guys in the gang get more violent, I don't have to decide too quickly."

"Yeah, probably best to pretend things are O.K. even if they aren't! You say you're launching your boat tonight?"

"Early morning tide if Lin machines the shaft today."

With that, they finished their meal and Hal walked her back to the office, hugged her, then set off for his truck.

All went routinely the next couple of weeks. Members of the gang directed nasty gestures and words at him as they passed him on the water, and Ray Look roared by him at high speed when he was fishing close to the ledges off Jocks Rocks, no doubt hoping to create enough disturbance to cause him to strike bottom. But nobody messed with his gear, as he feared they might. Hal spent both days and nights reviewing his life history and weighing choices. It was almost June, and the fishing was fair. He was tending 175 to 200 traps, checking half of them every other day, enjoying the freedom and beauty of the sea as they moved toward summer season. He got a few week-end orders from the newly arriving tourists, his dad's lifelong friends, where he could charge the full retail price. And he kept his gear close to the "territories" his family had more or less claimed during decades of fishing. He kept his eyes and ears wide open.

While fishing, he felt reasonably comfortable. But on the wharf, he was ignored or insulted. He even stopped attending Fishermen's Association meetings but learned through Randy, the informer who seemed to enjoy telling tales over the phone, that some members of the gang planned to move formal action in June to throw him out of the Association even if he paid his dues. That might give him trouble since only Association members could keep their bait barrels on Association property. But Hal decided to say nothing about it until the next session.

Late afternoons were rough for Hal. For years he had spent that time helping his father build more traps. A zillion times he'd heard, "God knows, we gotta do that!" The sea claimed so many that replacing them was a constant chore. In recent years he'd been a party to arguments that wire was better than oak slats for making traps. He had held to the traditional family claim that oak was the better though he had succumbed to the temptation to try wire and was fishing six, oak-reinforced traps to see how they fared. But now he didn't know quite what to do, musing, "If I stop fishing, move to Portland or wherever, there's not much point in having too many extra traps to sell...though I'm pretty sure the Barnsville guys will buy them even if Baysville fishermen boycott me."

Evenings and nights also wore heavily on Hal. There were always books to keep since tax deadlines were inevitable. He'd managed to get a Statesville accountant to make sense of his and his dead parent's April returns, but he wasn't so sure if he'd be so "lucky" the next year. And he continued to find himself sketching and daydreaming. His graceful "8's" and "V's" sometimes grew from a quarter inch to a full page in height if he felt that a sailboat looked more graceful embodied in such figures. His "L's" frequently became lighthouses along the horizon. In fact, he might begin an evening with a serious intention to work systematically on his books. But by the time he listened to the top-of-the hour news at 9 or NOAA Radio for weather forecasts, he would find himself filling page after page with sketches, much as he did when the teacher caught him in school. Sometimes while fishing, he'd stop to doodle on a new lobster trap or see the lines for a new sketch in the whirling water. Hence, he continued to wonder if he should forget fishing and go into some sort of commercial art.   He even wondered about going to college to become an art teacher. "God knows!" he exclaimed to himself, "Somebody needs to encourage kids and not destroy their natural talent."

One evening while brooding over this dilemma, Sue phoned, reporting that she'd had a difficult time reaching her mother who'd been vacationing on her yacht at Catalina Island.

"The news isn't good, Hal. She's determined that you carry on the family tradition. I'm not wholly sure why, but she certainly resented my asking her to 'cease and desist.'   Maybe she didn't like my new legal terminology. She said she'd write to you about it but gave me the permission to relay her 'No! No! a resounding No!' I'm sorry, Hal, for I think you should be free to do what you wanta do."

"Thanks, Sue, I'm sure you did all you could."

"She'll probably write as she promised. But just for kicks, she's not happy that we talked. And she doesn't like it because I'm part of the pro-abortion movement. She's a Phyllis Schafley follower. Know her?"

"Oh my gawd! Yeah, I've seen her on the news."

"Funny isn't it, how Neanderthal mom is, prob'bly carrying on that part of our family tradition! I'm sure glad I don't live in California where I might be under her thumb."

"Sue, you have my vote. Keep on keeping on, no matter what she says.   I'm O.K. for the moment. Things here are calm, not too much rough stuff these days, and I'll probably stick it out for the summer, at least. It's a nice season to be on the water. I'm working at my drawing. Hey! Come to think of it, would you like to see some?"

"Of course, send them along.   I may know somebody who knows somebody who would be interested."

"Great! I'll put some in the mail."

With that the cousins said good-bye, and Hal leaned back in his chair, pleased that he had confided in Sue at a crucial time.

Three days later, when he finally got to the Post Office to send his sketches to Sue, he discovered two letters and lots of junk mail in his box. He took one look at the return addresses, a formal-looking one from the Fishermen's Association and one on colorful and sweet smelling note paper from his Aunt Sarah. He decided to wait until he got home before opening either. But he fumed all the way down the road, sensing that neither of them contained good news.

Fixing himself a cup of coffee in the micro-oven, he sat in his favorite chair and read his Aunt Sarah's letter first.

Dear Nephew,

Sue called me last week to report on your luncheon conversation. As she told you, she pled with me to "cease and desist" (her words...the influence of her job in the law firm?) in pressuring you to keep your word on continuing the family's fishing tradition. This disappoints me, Hal, because I thought from our discussion at the funeral that you were sufficiently confident in your ability to keep David's and Gramps' business going. Now that I am 3000 miles away, this nonsense about your art and poetry crops its ugly head again. When are you going to grow up, forget such childish things? OF COURSE I know that fishing is hard and risky business. Didn't I grow up with it to watch your grandfather struggle against weather, the Great Depression, angry lobster gangs who couldn't tolerate his success. Why do you think I'm so far away from Maine? enjoying the freedom of my husband's wealth?

Because I was damn sick of the scrimp and save life your grandmother lived all the days of my childhood and beyond!

And why do I insist that you fulfill your promise to me? Because the Daniels family has a proud tradition to uphold, something that can't be said about many families that moved to America during the last century. In many ways, Hal, you are the last roots I have in New England, you and Sue. I'm happy in California, but I feel good about knowing you are there. I don't want to get nasty and push my threat about property and law-yers to the limits. As you know, I don't need the money, but I know you do. And sure as hell I don't trust lawyers!

I'm sorry you brought Sue into this before writing or phoning me. Sue has enough on her mind already, returning to the State where she spent college summers and getting adjusted to a new job. Maybe I shouldn't have gotten so angry with her over the phone? And probably the two of you do need to talk occasionally. I really do want you both to be happy. But I question how fair it is to put her between you and me.

When you've had an opportunity to digest this letter, perhaps you can phone me and we can talk some more. I can be reached most mornings between 9 and 11, 12 and 2 your time. You have my number. Call collect if you wish.

Love,

Aunt Sara

After reading the letter, carefully folding it and returning it to its envelope, Hal stared through the window at the lobster traps stacked in his back yard. Then, sort of snapping to attention, he opened the second letter. It was a form letter with his name and dates filled in:

> To: Harold Daniels
> From: President, Baysville Harbor Fishing Association
> Re: Attendance at Semi-monthly Meetings
> It has come to my attention that you have not attended
> the last   __3__   regular meetings of the
> Association. As you know, you face automatic
> suspension or expulsion if you do not attend the next
> meeting, to be held on __June 21__. Either suspension
> or expulsion will mean denial of your use of the
> Association's wharf and baithouse, also use of our
> meeting facilities and all other privileges attended to
> membership.
>
> > Very truly yours,
> > Joe "Digger" Danes

This time Hal slammed the letter onto the end table beside his chair, cursing, "Damn them! _Damn_ them! They don't miss a trick, do they?" He also knew that Digger Danes was one of the people who made the formal protest the morning after his boat was messed up. Too, Randy Ballou had warned him by phone that Digger had promised some of his pals, "confidentially," that he would "get Hal."

"Two pieces of 'good news' the same day," Hal mused, automatically putting on his jacket and sauntering to the shed to work on traps. He knew it was too early to give up. "Maybe," he muttered half aloud, "they want to make a fighter out of me."

That night he tossed and turned for hours. He weighed his options over and over until he felt his head would burst. Finally he got up about four o'clock, dressed, and went to the wharf, determined to continue his work no matter the odds. As he slipped into his skiff, he heard voices on the water in the direction of his boat. He recognized one of them as Walt Carmen's, but not the other. He noiselessly rowed himself toward his mooring, his heart racing as though to burst his chest. He whispered to himself, "I mustn't panic or I'll warn 'em." He knew they'd as soon beat him up as eat breakfast. It was pitch black, shortly before dawn. And the men in the other boat ahead of him continued to laugh and chatter as though going to a picnic. He was sure they were up to no good and only wished he had brought a flashlight to catch them in the act of doing their mischief. Resting on his oars and gliding, Hal heard them reach his boat; heard the hiss-s-s of a spray can, then knew they were fiddling with his mooring hawser. He also heard a gnawing sound so figured they must be cutting his boat free.

"Now," he mused, "it's time to scare hell out of 'em!"

Familiar with every sound possible around a dock, he tapped his oars in the oarlocks just enough to set up an echo against the buildings at the wharf. He then heard the men in the other boat turn quickly, knocking their skiff against the side of his boat, swearing, "Let's get to hell out of here!" They rowed off as though the devil were after them. Now he knew that the second voice was Digger's. Hal sat quietly in his skiff and saw them row swiftly across the thin water reflection of a distant yellow light. While he could not recognize their forms, he noted that there were only two men in the other skiff. When he saw them row for the open sea rather than the wharf, he sensed they were confused and frightened so he moved quickly toward the "Mary Q." Daylight would arrive in a half hour or so, hence he devised a scheme to scare them some more.

Without starting his engine, he went immediately into the cabin to get a gallon of white paint, grabbed a brush, then prepared to start up the "Mary Q" as soon as the first streaks of light appeared. "If," he reasoned, "they did spray the side of the boat, I'll paint it quickly, then roar out of the harbor and go by them on the newly painted side." Meanwhile, he climbed atop the cabin to check the rope; sure enough they'd slashed through one of the strands, evidently intending to set his boat adrift and let it bang into the other vessels, causing whatever damage wind and tides might direct.

When the first light of dawn rose from the east, turning it saffron, Hal was in the skiff and painting over the red paint the two men had used to 'decorate' his boat. Slopping on white paint vigorously, within a couple of minutes he'd started his motor and was headed for the channel leading toward the open sea. He knew the men in the boat really had no nearby refuge so must be "out there somewhere" to avoid getting caught at the dock. Sure enough, he soon spied a black speck on the water, the two men rowing vigorously toward Barnsville. Hal was unfamiliar with the bottom in that direction so he pulled his charts and calculated how close he wanted to be when he passed the two men in the small boat. He also hoped they were not carrying guns, but decided to risk that possibility. Steering directly toward their skiff, Hal set the throttle at full speed and bore down on the two men. Obviously they saw him coming and began rowing in another direction. Since Hal could maneuver more quickly than they, he checked his chart and simply turned his rudder to follow them. He was determined, too, to show them the side of the boat that they'd painted, hoped they'd treat it like a ghost. Within a hundred feet or so, he cut his throttle and steered between them and the shore, sort of drifting past. He quickly recognized Walt, then sure as hell, there was Digger.

When about twenty feet away, Hal said, very pleasantly, "Good morning, Walt; hi, Digger. See you sometime back at the wharf." He said no more, but he saw them stare at the can of white paint that he had deliberately left on the stern, the brush laid across it. Allowing time for the "Mary Q" to drift on its own momentum, for he didn't want to swamp the two men much as he might loathe them, he turned sharply toward the open sea, opened the throttle and headed east toward Jocks Rocks. As he peered back at the two men, saw them turn around and row for Baysville Harbor, he chuckled to himself and wondered if he should have used his Polaroid camera to record the scene.

Throughout the morning as he thought about what Digger and Walt had done, he worked himself into a tizzy. And the angrier he got, the clearer his resolve became. He made decisions over which he'd merely brooded before. "I'll stay and fight," he declared, shouting over the sound of the whinnying winch. "I'll double...I'll triple my traps and hire a sternman strong enough to help me with the physical work and be loyal enough to improve my confidence. I'll show these bastards that they can't drive a Daniels from his boat and livelihood. And I'll get so paranoid, they'll get away with nothing."

He also dialogued in his mind about how these two inner members of the gang would behave when he walked into the June 21st Fishermen's meeting. Too, he resolved that as much as he hated guns, he would buy a rifle and a pistol, take lessons in using them, and practice by the house in plain sight of the road, then mused, "Hope I don't have to shoot somebody before they'll leave me alone."

He further promised himself that he would phone both Sue and his Aunt Sarah the next Sunday, to tell them of his decisions.

Chapter 3

More Stormy Water

While Hal had, indeed, done much daydreaming throughout his life, the practical training under his parent and grandparent had inculcated the values and satisfactions that so often result simply by "getting things done," sometimes semi-automatically. Not only was it a family tradition, as his elders had taught him, but it was also a Yankee virtue as they knew it.

So during the week following his decisions on that fateful morning of "discovery," Hal continued to repair old traps and build new ones with deliberate speed. He aimed to have at least 300 built and set by the Fourth of July, a date that had served as a landmark in his family since Andrew Jackson was president. His great great grandfather, for instance, always made a point to pay up all of his winter bills by that day, a ritual carefully recorded in his diaries. His great grandfather always took the family to Statesville for the parade and fireworks. And his grandfather always wanted fresh peas by that day. Hal also phoned Sue and his Aunt Sarah to tell them of his decision as well as announce that he was entering a drawing contest in Statesville. He wanted to deliver his aunt a defiant message, that he wasn't being entirely emasculated. Also, after years of hesitation, he was pleased to have made the decision to enter the contest, "devil take the hindermost."

His relations with those on the docks and ocean did not improve, but nobody tried any "funny business" on him. He followed through with his plan to buy guns and learn to shoot accurately and responsibly, taking special pains to target practice in his back yard where he had plenty of open space and could be seen by all those who drove past the house.

When it came to adding a sternman, he knew that it probably was impossible to find anybody in Baysville Harbor who would dare work with him. So he decided to tell old family friends in Barnsville and Statesville about his needs. He figured that some one of the high school or college kids might like to work a few weeks during the summer, earn some money and perhaps "learn the trade." Meanwhile, he could begin to look for somebody who might want to join him on a long-term sharing basis. One day when the sea was rough and thunder showers threatened, he drove the nine miles to Barnsville, found his grandfather's close friend, Jerry Moulton, and got down to "hard facts and brass tacks." Jerry boasted of having "cut his teeth" with Gramps so welcomed the chance to talk.

"Sorry 'bout your grandfather and dad, Hal. No bettuh fishermen on the Maine Coast. If you folluh in their footsteps, you don't have nuthin' to worry 'bout."

"Thanks, Jerry, I need friends like you... badly."

"Yeah, heard they sorta tried to scare you, almost phoned to encourage you to 'hang in theyuh.' Got doin' othuh things. Glad ta see that you didn't give up. Some of them fellahs are pretty rough and tumble. I got inta a pissin' contest with them when 'Digger' Danes crossed over the line and set his traps where he knew I'd worked the bottom for years. Threatened me with his gun one day so I rallied my fishin' buddies. We gave him a lesson he'll prob'bly never forget. None us of evah talked 'bout it very much. Just a warnin'...."

"Did it come to shootin', Jerry?"

"Nope, but we expected him to cut our gear some dark night. But he musta got the message and nevuh came back. Really a cowuhd at haht."

"Look, Jerry, I came over to ask for help. I'm gonna expand my gear, may fish as many as 500 traps in the next couple of years. Need a good sternman. Know anybody lookin' fer a job? Felluh home from college or a strong high school lad?"

Jerry turned his head and stared long and intensely toward the docks, finally weighing his words, "None of my business, Hal, and you can do as yer please, but it's a tough time to be expanding. Watuh's already full of traps, days when it looks like you could walk around Jocks Rocks or across Barnsville Harbor hopping from buoy to buoy. Prices ain't good. You could expand yourself outa business."

"Yeah, I know, Jerry, but at one time dad and Gramps fished 500 to 700 traps between them, I figured there was room."

"Maybe yes, Hal, maybe not," Jerry responded, setting his jaw squarely against a fresh, salt-laden wind blowing from the harbor.

Hal said nothing but peered quizzically into the elder's face. "Well, don' mean to put the pressure on ya, Jerry. Jus' thought you might know somebody. If yer do, will yer gimme a call? You know my phone number."

"Ayuh, glad to, Hal. And don't get upset 'bout my viewpoint. I just don' wanta see ya go undah. Bad time for lobstermen. Too many problems comin' from the outside. You know that!"

Hal reached out to shake his friend's hand, remarking, "Yup and it's O.K., Jerry, I understand. Thanks for bein' frank. No friend's gonna tell another friend only what he wants ta heeah." He got into his truck and headed back to Baysville.

While driving along the crooked wooded road, he had another idea. He would put an ad for a sternman in the Statesville GAZETTE. He'd seen such ads in the local paper, knew such an ad might get lost in the Portland PRESS HERALD or Sunday TELEGRAM. So he began composing the ad in his head as he drove along, negotiating the sharp bends. By the time he reached the house, he had it pretty much in mind, so grabbed a pencil to jot it down:

> WANTED. RELIABLE STERNMAN TO ASSIST
> DESCENDANT OF VETERAN LOBSTERING
> FAMILY. DIFFICULT HOURS BUT GOOD
> REWARDS. APPLY IMMEDIATELY BY WRITING
> TO BOX X-77...or PHONING 777-llll.

As he finished the ad and dialed the GAZETTE, he recalled how he always admired Gramps for getting a PO Box, lobster licenses and phone numbers he could remember. He quickly read the ad over the phone, gave his name and address, was assured the ad would appear in the next three days' editions of the paper.

Hal spent that afternoon building traps and the evening sorting through a stack of sketches extending back to high school. He made a note on his "To Do" list to take a trip to Portland to buy more art paper, following the specifications the Whaleback Gallery recommended for submissions to the drawing contest.

Another week passed. All went well. Hal fished mornings, built traps afternoons, worked on his sketches evenings, got to bed by nine so he could rise at four or five and be on the water near daybreak. The routine felt good.

Two nights after his ad for a sternman appeared he got a phone call at 2:14 a.m. He stared in disbelief at the red-eyed digital clock beside the phone, lifted the receiver with some apprehension. A high-pitched voice simulating a woman's said, "Hello, dearie, I'll be your sternman if you'll pay me enough," then hung up.

Hal charged it off as a crank call and soon drifted off to sleep. Again the phone rang. This time it was 3:03 and it really rattled him. A deep male voice said, "You don't need no sternman, you need a hole in the head!"

By this time Hal was fully awake so had presence of mind enough to call the operator to ask if there were any way to trace calls. She told him to phone their customer representative the next morning. And he knew it was useless to call the police. So he decided to get dressed, stay up, and see what happened. Sure enough, he got three more calls before his alarm went off. But he'd heard that blowing a whistle into the phone might dissuade cranks. Hence he was prepared for the fifth call, waiting until the disguised voice started, then blew a whistle as loudly as he could. Immediately he recognized Walt Carmen's voice, "You sonavabitch," then "Click!" Hal made a mental note to see Walt that day to check his voice recollections. That morning he bided his time until Walt was in the baithouse before saying, as pleasantly as he could, "Good mornin'."

Since Walt did not reply immediately, Hal repeated his greeting. Walt muttered something which Hal didn't understand so he asked, "What say?"

"Oh, nuthin', nuthin'!"

Hal was quite certain that the last crank call was Walt's, but he held his tongue when he heard others in the gang clomping along the wharf toward the baitshack.

As Hal rolled his barrels toward the winch at the edge of the wharf, Digger confronted him with a snarl, "Heah ya lookin' for a sternman?"

"Yup!"

"Bettuh not, gonna be trouble."

"Why?"

"Nuf fishin' out thayuh already."

"Well, we'll see, Diggah, we'll see," Hal responded and rolled his barrel into its usual place, anticipating more nasty remarks from members of the gang. To his surprise, they were quiet. But when he pulled the "Mary Q" up to the dock, he discovered that somebody had tipped one barrel of bait onto the wharf, forcing him to put it back in the barrel with a pitch fork. As he cleaned up the mess and hosed the dock, he heard Digger, Walt and Randy snickering among themselves; he knew they were up to no good. Finally, he saw Randy approaching him. "What now?" Hal thought.

Randy began, "Been talkin' with the boys. We agree you gotta get outa Jocks Rocks. Do that and they'll call off the dogs."

Hal was angry, but he knew he had to control his words. He paused for a moment, looked toward the harbor channel, finally replied, "My family's been fishin' theyuh for six or eight generations, Randy. Room enuf for all of us. So why don't you guys find your own bottoms, fish theyuh, and leave me alone? I ain't int'rested in a fight unless you guys force me into it. And hey, tell your friends that paintin' up a storm on the side of my boat ain't gonna get 'em anywhere either. Tell 'em I said so."

Hal knew that for all of Randy's bravado he was just a pawn in Digger's hand and would do nothing by himself. He owed them too much. So Hal watched from the corner of his eyes, heard the group whispering and laughing among themselves. He also was aware that this was the night of the Fishermen's Association meeting, and he didn't intend to be intimidated into staying away. When he got his bait aboard and headed for the open ocean, he looked back to see them giving him the finger. Tired as he was from having been up answering the phone, he did not respond in kind. His morning difficulties blended with pleasant surprises at finding so many lobsters in his traps. But out of curiosity he turned on his CB to tune into the gang's raucous and profane banter. He had a feeling that they would reveal their plans via the airwaves because they knew he rarely listened to their chatter. Sure enough, there was Ray Look jiving with Digger.

"He won't show up t'nite, Digger. We'll be rid of him."

"Hope ya right. I'm tired of the little squirt's threats."

Walt cut in, "What gives him the right ta claim his dad's fishin' grounds, best in our stretch of coast."

"Oughta be a law agin it!" Ray exclaimed.

Digger chimed in to say, "Hey, there's an ideeuh. If we can't drive him off, maybe our Association can get our rep ta sponsuh such a law in 'Gusta?"

"Take too long," Walt remarked, nervously observing, "Let's shoot the bastahd and get it ovah with."

"Easy, Walt, Elmah Kruse knows 'bout the bad blood 'tween us and the Daniels. I don' want no time in prison."

"I don' think anybody'd know who done it," Ray exclaimed. "His boat could explode accidentally, and who'd know the difference?"

Digger warned, "Yeah, but 'member he now has guns; rumor has it that he's a good shot."

About that time Walt's wife cut into the discussion, "Walt...Walt...Walt....!"

"Yeah, Alice, what's th' mattah?"

"Dang'rous talk, Walt, bettah stop it. Problems heah."

"What's the mattah, hon?"

"Skunk under the kitchen. Get back heah fast!"

"But I just began ta fish!"

"Yeah, I know, but you can fish tomorruh. I need help now ta get this critter outa heah."

Walt turned off his CB and Hal guessed he was probably swearing. He couldn't help chuckling as he reflected, "Leaving one group of skunks to tend to another!"

Looking across the open expanse of sea, Hal saw him do a U-turn and head for the harbor. Hal had heard enough to know that he must keep up his guard. Finished with tending his traps by noon and hearing the Barnsville whistle blow, he knew the wind had come up and he'd better get home to see if he'd gotten any responses to his ad.

Back at the house Hal found a letter under his doormat, one which intrigued him:

Dear Hal,

I hope you will remember me as the girl in your class who protested when our classmates took all those negative votes about your promises of success. I remember your drawings, too.

I just read your ad in the GAZETTE and would like very much to talk with you about the job. Since I last saw you at our graduation, I have served as a clerk at K-Mart, commuted to Portland to take courses to become a Practical Nurse, and now work in a local nursing home for senior citizens. None of those jobs has excited me. I still live with my parents, but they're getting as fed up w ith me as I am with them...and myself. I liked outdoor sports in high school and now play in a softball league. Maybe I was cut out to be a lobsterwoman? I know there are only a handful in the State. Could we talk? Incidentally, I got your name and address by looking up all of the Baysville Harbor numbers in the phone book. After you've had an opportunity to read my note and consider the complications that might come from having a woman aboard your boat, would you please phone me? I know you fish days, and I work nights. But I don't go to work until 7 o'clock so you could reach me around supper time.

Hoping you'll be willing to talk, I am

Sincerely,
Maria LaRoux

When Hal finished reading the letter, he carefully folded it and stared for a moment toward the dunes between his home and the beach. He watched the wind tear at the grass heads, bending them nearly into the sand. He heard rumblings, turned to look toward the west to see heavy thunderheads advancing. He made a mental check of his boat, reminding himself that he had everything secured, the cabin door closed, and nothing on deck that could blow into the harbor. He then looked back to the stack of drawings he was sorting on the dining room table and began imagining Maria as a sternman or sternwoman. He wasn't even sure how to say it.

He knew his dad and Gramps would have said: "Nix!" because of superstition about women on boats. He figured, too, that the gang would harass him the more. Nor was he sure that he wanted an additional distraction, for he had a clear recollection of Maria's lovely face at graduation when she offered him the only support of the evening, other than his parents'. Also he wondered, "I've not seen her for three years; I wonder if she'd be strong enough to wrestle with bait barrels, traps, gear, the anchor, oars on the skiff...to say nothing about handling guns."

In the midst of this musing the phone rang; it was a high school fellow Jerry Moulton recommended, wanting to know if he could be interviewed for the job. Hal responded politely, "I can see you tomorrow afternoon at 3:30 if that's O.K. with you?"

"Fine." When Hal hung up, he thought about the fellow's being courteous; then said to himself, "If Maria can make it only at 4 tomorrow, maybe I've goofed?"

Somewhat nervously he fixed an early supper, ate standing up, and at exactly six rang Maria's number. Her father answered, in French. This threw Hal for a moment, but he finally stammered, "Is-is-is-s Maria there?"

The father shifted to English, and snarled, "Who's this, Mr. Jacques at the Home?"

"No, Mr. LaRoux, I'm Hal Daniels, one of Maria's high school classmates. May I please talk with Maria?"

This calmed the man a little, and he shifted back to French, "*Une moment!!*" There followed what seemed like "the longest pause" in Hal's life, and he could hear what he supposed was Maria's mother and father, conversing in high-pitched voices, speaking French. He only caught an occasional word, but heard his name mentioned several times. Finally, Maria reached the phone, seeming very sleepy.

"Hel-lo."

"Maria? This is Hal Daniels."

"Yes, Hal, sorry to take so long to get here. I'd overslept."

"Maria, I got your letter. Want me to call you back?"

"No, no, this is O.K. I should have been up and awake by now anyway. Glad you caught me before I left for work."

"Look, Maria, I can't promise anything over the phone, but I would love to talk with you."

"I'm glad, Hal," Maria practically bubbled. "Where shall we meet?"

"How about McDonald's in Statesville tomorrow evening? Isn't that fairly close to where you work?"

"A couple of streets away. Great! What time?"

"I've gotta be here at 3:30, but could make it by 5:30."

"Terrific, Hal. It'll be good to see you."

"And you, too, Maria.... Bye."

Twenty minutes later he left for the Fishermen's Association Meeting, fully braced for what might happen. Tempted to take his pistol, he decided against it at the last minute, musing, "Better not make it worse than it is."

As he walked into the meeting, Digger's letter carefully tucked into his shirt pocket, Ray Look, half stewed, greeted him at the door, "Ev'nin', Hal, come aboard and get a hake!"

The comment was hardly shocking. Hal had heard it since a kid, usually in situations calling for humor. Nobody in Baysville Harbor had caught a hake for years!

Hal acknowledged the greeting, strolled casually toward the chairs in the half-filled room, sat in the back row, musing, "I want to observe the entire scene, I don't trust my back to any of these characters who want to drive me off Jocks Rocks."

When the meeting finally opened, Digger Danes in the Chair asked the secretary to pass out the agenda. By the time he reached Hal, he'd run out of copies so returned to the front table. When Digger asked if everybody had a copy, Hal stood up, said he didn't and asked, "Can anybody share theirs?"

All twenty-seven fishermen present turned toward Hal and began muttering to one another. Hal could see the anxiety in their faces, but nobody offered to share their copy of the agenda until Ray Look, now fully inebriated, staggered back along the middle aisle to hand his to Hal midst cat calls and boos. Ray then teetered and swayed back to his chair. Digger banged the gavel to quiet the room. Then, without the slightest hesitation he said, "If you cause any further commotion, Hal, I'll have Elmer remove you from the meeting." And he stared sternly at Elmer Kruse who just "happened" to be present.

Hal said nothing, studying the agenda to see what further mischief Digger and his gang had concocted. Sure enough, among standard items of Old and New Business was one labeled, "Harold Daniels - Membership." Hal braced himself.

Most items on the agenda were fairly routine: approval of the minutes, status of the budget, discussion of renegotiating the loan for repairing the docks some years before. They were unanimous in passing resolutions supporting uniform gauging and fishing seasons for Canadian and Maine lobsters, also for condemning yachtsmen who ran over their buoys in the harbor channel and draggers who messed up their gear outside. But they couldn't reach agreement on limiting trap numbers. Some wanted to follow the lead of fishing communities down state who voluntarily limited their numbers; others were "agin" it. All were in favor of writing to the Maine Congressional delegation to keep Soviet fishing boats "as far away as Moscow!" Others felt that the U.S. Navy and Coast Guard took too many liberties running over their traps. But there was mixed opinion on whether or not to boycott marketmen who cut prices in the forthcoming summer, hence they tabled the item.

When they got to membership items, the group approved taking in a new sternman working with Laurence Bigger as well as applauding his entry into the Association. Hal watched carefully as they approached the agenda item concerning him. When the moment arrived, he was prepared. He unfolded Digger's letter warning him to attend that night or be thrown out for missing three meetings in a row.

Digger introduced the item by giving a sarcastic speech about the way Hal's grandfather and father had comported themselves during their lives, run roughshod over all of them sitting in that room and stressed the way they monopolized the territory around Jocks Rocks. "And now," Digger went on, "this young punk, more famous for his drawing and poetry than his ability to run a boat, has moved into what he thinks is his family's territory and done it arrogantly. Does he think we're stupid and will stand aside quietly while he fishes in the best grounds around Baysville Hahbuh?" He concluded by reading his letter to Hal, then remarking, "Technically, Hal is here tonight so we can't throw him out of the Association on membership rules alone. But twice now, once at the dock last month and just a few minutes ago, he has disturbed the peace of Baysville. I thought this might happen. That's why I asked Deputy Kruse to come to tonight's meeting. You all saw it for yourselves. Now, some of our members have urged me to bring these troubling matters to your attention. We simply will not tolerate this young whipper-snapper upsetting our lives. Do I hear a motion recommending that we cancel his membership?"

Walt Carmen moved, and a couple of mumbled "Seconds" rose from the floor. Digger queried, "Discussion?"

The room was silent until Hal stood, walked slowly to the front of the room, looked Digger straight in the eye, then spoke calmly but with deliberation. "Digger, you know as well as I that all of these charges are as phoney as a left-handed monkey wrench. You don't like it because Jocks Rocks is the most productive section on this part of the Coast. You know as well as I who painted my boat when I had it out of water repairing the shaft and wheel. You also know how many copies of the agenda you need for these meetings. Also you know that I have no fame as an artist and poet, just a home-grown reputation for being infamous...in YOUR eyes!"

Hal then walked back and stood beside Walt Carmen who had been with Digger the morning they painted his gunnels before dawn, continuing to speak, "Now, Digger, maybe you and Walt would like to tell the group what you did to my boat just before dawn recently, then how the two of you tried to avoid detection when I caught you in the act. And maybe you'd like for me to show you the infrared pictures of that scene? Maybe Deputy Kruse would like to see the pictures too? or if I should decide to show them to Sheriff Owen, maybe he would take an interest in this little war you've started...to punish me because of your jealousy of my family's skills as fishermen and your contempt for my life history!"

Walt shuffled in his chair, stood as though to go after Hal's throat, but Warren Reed put a firm hand on his shoulder pulling him back into his seat.

Hal saw the brief scuffle and while he was basically shy and felt butterflies dancing in his stomach, he stood his ground. He knew he was bluffing about the infrared pictures, but hadn't forgotten what he'd learned in high school speech classes. So he stared for a moment at Walt then calmly walked back to the front of the room, addressing the whole group again. "I think some of you would be my friends if some others of you were not so intimidating. I'm young so you think you can scare me. You've teased me because of my drawings and my poetry even though only one or two of you have even seen either. I admit I screwed up when crashing boats in the harbor last winter. But you won't scare me away because you're too frightened yourselves. And let me tell you, if you vote me out of this organization, every single one of you is a sitting duck to getting the same treatment...if you don't do what the leaders say. And none of you quiet ones with families can afford to be destroyed if or when these warlike tactics are investigated by county authorities. So I would recommend that you defeat the motion to ban me from this group and do it by closed ballot."

As he ambled slowly back to his seat, a few of the men applauded...until they looked up to see Digger and other officers glowering. Hal noted that Elmer Kruse appeared nervous but was clapping his hands. He found some consolation in that. "Maybe," Hal thought, "he's not in Digger's hip pocket?"

After Hal sat down, the men chattered animatedly among themselves, and Hal hoped this would not be another strike against him as Digger began banging the gavel. When things quieted down, Jack Small asked for the floor. "No question in my mind," Jack began, "that things here in Baysville have gotten completely out of hand, and I think all of these strong-arm tactics should stop. Why should we as a group punish Hal Daniels, even if we didn't like his family and the power they wielded? Somebody's going to get killed if this skirmish doesn't stop. Hal's father and grandfather and great grandfather fished Jocks Rocks and did it well, gave us competition we sometimes couldn't take. And without those men we'd never have had the guts to borrow money to build a dock. They developed the market which we use. Let's let by-gones be by-gones. We should be expressing sympathy for Hal about his family's loss...not trying to destroy him. I would like to ask the person making the motion to have Hal Daniels removed from this Association, ask him to withdraw his motion. It never should have been made."

Again the men chattered noisily when Jack sat down. And again Digger banged the gavel to restore order.

This time Elmer Kruse asked if he could speak though he was not a member of the Association. Digger grimaced a little but gave him the floor. "Naturally, I know most things that have happened here in Baysville since Hal's father and grandfather died. I also know there's been a lot of dirty business, but I won't reveal the names of those who've tried to bribe me to be tough on Hal. They know I did not take a dime. Nor will I say what I know about the paint-spraying incidents. But I will say you'd damn well better stop blaming the victim for what's been goin' on here the past couple of months....or else...you'll all be victims if Sheriff Owen comes in here to investigate and take action...and he tells me he will do just that if you keep it up. I think that Jack Small has the right idea, take the motion off the floor, expunge the records, and get back to fishing."

Warren Reed got the floor next, said simply, "I agree with Jack Small. Let's get back to fishing and stop this nonsense!"

More noisy discussion. Until Digger gaveled for silence. Turning to the men who made and seconded the motion, he asked, "Are you willing to withdraw the motion?"

Neither would move an inch. Digger, sensing that the mood of the group had turned in Hal's direction, observed somewhat sarcastically, "Since it would seem that everybody wants to save our young heroic poet-wimp, we'd better have a secret ballot. Elmer will you please distribute these pieces of paper, collect and count the votes?"

Elmer agreed. Digger scowled and looked pointedly at Hal, remarking, "And the chair rules that Hal Daniels will not vote, a clear case of conflict of interest."

During the balloting Jack Small handed his vote to Deputy Kruse then walked back to talk with Hal. After shaking hands, he said, "Look, Hal, I'm awfully sorry to be silent all this time, but as you may know Wife Nettie's had an operation and has been awfully sick so I've been fishing only once a week, as you may have noticed. As I've told you before, I liked your father and Gramps. I'm glad you're still fishing. And you deserve hanging by this lynch gang like I want a bomb in my boat. I think the vote will go your way. Meanwhile, if there's any time you need a friend, you know where I live."

"Thanks, Jack, I may call on you."

Warren Reed also came along with a similar explanation for not having stood by Hal. "Spent most of my time in the hospital with my son-in-law."

"I've missed you, Warren, many thanks."

After Deputy Kruse handed the results to Digger, he banged the gavel hard enough to crack the table. He was clearly rattled as he coughed, spit into his handkerchief and announced, "On the motion to ban Harold Daniels from this Association, the vote is 6 Yea and 20 Nay and one messed-up ballot, probably Ray's, so the Nays have it."

A cheer went up from the group; Digger immediately declared "Adjournment" and several dashed back to shake Hal's hand, most expressing shame or sorrow for the entire affair. One member waved his hand vigorously and shouted, trying to get Digger's attention about violating Roberts RULES on adjournment, but the majority were anxious to get out of there!

Hal left the building in better spirits than when he came, but he could not overlook the fact that six people sought his ouster.   Although he was not sure of the identity of all six, he knew four who would bear watching.

Chapter 4

Stern Fellows and Sisters

When Hal got home that night, he felt both wound up and letdown. Deciding to do nothing serious and watch a TV film, he set his alarm an hour later than usual to get some much needed sleep.

He was hardly into the film when the phone rang. It was Jerry Moulton. "Tried all evening to get you, Hal, to see if my nephew had phoned about the sternman job."

"Out to Fishermen's meeting."

"I figured."

"Nephew's name, Jerry?"

"Ed Taylor, my sister's grandson."

"Yeah, he called, gonna see him tomorrow afternoon."

"Good boy, but lay out all the pros and cons. Ain't had much experience, still wet behind the ears. Don' be 'fraid to turn him down, Hal, if he's not right for you. I'll understand."

"Thanks, Jerry."

When Hal hit the floor the next morning, the bright sun hurt his eyes. But he pinched himself as a reminder that some of his woes seemed to be behind him.

Yet, he continued to repeat, "Watch out for the six, six... Can't let down my guard."

By the time he reached the dock most of his friends and enemies were already out fishing. Even Warren Reed who'd been away at the hospital so much of the past couple of months. But Hal concentrated on his work, moved routinely to the dock for bait then on out to Jocks Rocks, enjoying the beauty of the morning, especially the fleet of lobsterboats lying off to the southeast and silhouetted against the sun-painted ocean. "Be prettier a little later," he thought, "when they break out their sails." He scribbled the first line of a haiku on his lobster tote pad: "Black-gold triangles..." Looking back to sea, he mused, "I'll finish it later..."

Pulling his traps, he was pleased by the catch, enjoyed watching the gulls follow in white noisy clouds, circle his boat and dive for crabs and bait fragments he tossed nonchalantly over the side while preparing each trap. One gull messed up the visor of his cap, reminding him of the hat a friend had given him, with woven simulation of shit all over it plus the caption: DAMNED SEAGULLS. He smiled as he thought of it, adjusted his sun glasses and looked up to watch the wheeling gulls hang on the breezes, glide to bomb his boat or dive for breakfast.

He debated for a moment about whether or not to test the CB network, finally could not resist temptation and flipped the switch even as he noticed his new sonar screen flickering strangely. As the voices began to crack and rattle over the CB speaker, he played with the nobs on his depth finder.

"Can't be," he muttered to himself; but upon peering overboard, he saw the water teaming with huge churning fish. About the same time he heard Randy Ballou excitedly exclaim, "Damn it, the water's swarming with cod. Can't believe it. Nobody's seen schools like this for years."

Jack Small cut in to exclaim, "Un-be-liev-able!"

Hal quickly checked his distance from the nearby ledges, then wheeled into deeper water, reaching for his fishing pole wound with 20-pound line and a shiny stainless steel sinker-hook. "Gonna get myself a mess of fish, no matter," he muttered as though somebody were listening. He observed others putting up their sails to steady their boats into wind and tide, turned off his motor and began to drift. Within seconds of casting his line overboard, he pulled in a cod which he estimated must weigh ten or twelve pounds. Looking up to check his drift, he saw others landing fish. Within twenty minutes he had fourteen. "What I can't eat I'll use for bait." Also he made a mental note to take a couple of fish to Jerry Moulton if the school didn't get down as far as Barnsville. His CB connections told him that his Baysville associates were having the same kind of luck, were excited and amazed! "What a difference," he thought, "from the conspiracy crap I've been facing. Maybe that's it," he conjured, "if the fishing's good, perhaps they'll take their resentments off my shoulders."

Back at the dock there was much excitement about the school of fish and their catches. Some of the fellows were rushing home to "fetch" their nets and planning to return to the sea that afternoon and to try for a big catch, one they might take to the Portland auction house or even to Gloucester. Doing that crossed Hal's mind, but he quickly shifted back to his plan for the afternoon: ice his catch, then meet Ed Taylor and Maria La Roux. On the way home he checked his mailbox and found one other inquiry about the sternman's job, one with two pages of questions, including one about retirement benefits. The fellow also enclosed his resumé, indicating that he attended an Ivy League College (he didn't say which) and asking for a return phone call between 5 and 5:10 the following day. Hal shook his head in disbelief as he read it, almost automatically deciding not to employ him. "Doesn't fit my profile," he muttered. But he folded the letter, stuck it in his shirt pocket and rolled his pick-up along toward home.

Deciding to have fried fish for lunch, he quickly filleted the largest cod, pulled a burlap bag and ice over the others, and carried the tub down into the cool cellar. While eating, his mind raced ahead to the two interviews. One thing he knew. He had to shower and change clothes to shed the smell of fishbait! It even occurred to him that he'd better dress up to see Maria, the first time he'd had such a thought since seeing Sue in Portland.

At exactly 3:30 he heard a knock at his porch door. It was Ed Taylor, a rather tall and thin fellow of 15 or 16, blond, tanned cheeks and a firm handshake. Hal greeted him, asked if he'd like iced tea or coffee. Ed chose tea and they went to the kitchen where Hal began asking questions.

"Your uncle didn't say why you wanted to be a sternman, Ed. Can you say something about that?"

"Yeah, I like the outdoors, play tennis all spring and summer, don't want an inside job."

"Have any idea what the job's like?"

"Some....been fishing a couple of times with Uncle Jerry."

"Do you know what the hardships are, Ed?"

"What do you mean?"

"Well," Hal continued, "it means getting up before dawn, heavy doses of salt air and sun, constant motion of the boat, handling smelly fishbait that even turns my stomach at times, lifting heavy baitbarrels and lobster traps. You must have seen some of this when you were fishing with your uncle?"

Ed looked off over the marsh, sipping his iced tea as he draped his long body into a chair and crossed his legs, finally responding, "Yeah, I suppose I did, but I went along for the ride so may have missed the details since they didn't apply to me."

"Ever been seasick, Ed?"

"Got a little queasy deep-sea fishing once, shucking mussels in a heavy roll. Combination of smell and looking down into the boat did it. Stood up, though, and it cleared."

Then Ed began asking questions. "How much will the job pay, Mr. Daniels?"

Hal interjected. "Call me 'Hal,' everybody else does and not that much age separates us."

"O.K., Hal."

"About money: I'd have to start you at about minimum wage, say $3.75 or so an hour 'cause the fishing isn't all that great now. When it picks up toward the end of the summer, I can increase it. Also you'd have more experience by then."

"I'm not complaining, Hal, but you know I can make almost double that at a McDonald's or one of the motels. There's a labor shortage along the coast most summers."

"Yeah, I know, but most of those jobs are inside and don't require the skills you'd learn as sternman."

"Good thought, but would the skills I learn here apply if I were to spend my life further Down East?"

"Some of 'em, Ed, on setting traps, maneuvering the boat, handling lobsters, etc. The ocean bottom varies from cove to cove; and, of course, your relationship with others in any lobster gang might be different Down East."

"The lobster gang?"

"Yeah, the men who work in a particular area. For instance I am identified as working with the Baysville Harbor fishermen, belong to their Association..."

Hal stopped in the middle of a sentence, looked out toward his boat, and his recent troubles and the previous night's meeting flashed before his eyes.

Ed caught the pause, observing, "Yes?"

"Sorry, Ed, I was just thinking about the Baysville group. It's O.K. But your uncle," he continued, " would be a member of the Barnsville gang."

"Uncle Jerry says they sometimes fight?"

"Sometimes, but I don't think you'd have to worry. Everything is calm here right now, and we get along pretty good with the gangs on either side of us."

"Have to admit to you, Hal, that I'm pretty much of a coward. I'd rather run away than fight."

"No problem, Ed. Them's my sentiments, too."

As Ed prepared to leave, he asked, "When will I know if I get the job? Will you phone my uncle and tell him because I'm just here for the summer and staying with a friend?"

"I should know by tomorrow or next day. Have a couple of other persons to interview. O.K.?"

"O.K."

As Ed slipped through the door and climbed into his classic VW bug, Hal thought to himself, "Sorta like the boy, but he doesn't seem too strong. Hope I didn't lie to him about trouble with the gang?"

At precisely 5:30 Hal pulled his truck into the Statesville McDonald's. He wasn't quite prepared for what he saw. Maria was standing at the corner door, near the kids' jungle jim, dressed in her nursing uniform. He had remembered her as being "cute," but had hardly remembered her as well built, five-five or so, 110 pounds with black boyish bobbed hair, modest breasts and a willing smile. Even before he left the truck and rushed to shake her hand, he was asking himself, "Golly, is she strong enough to do the job?"

Formal greetings behind them, they found a seat in a quiet corner and Hal did the traditional thing, asking her what she would like to eat and drink. Right off, Maria said, "You know, Hal, I'm a feminist, usually serve myself."

"That's O.K. with me, Maria. My mom and dad taught me to respect women and be polite. I don't have many to respect since grandma and mom died. I'm prob'bly not up to date."

Maria reached across the table, put her hand on Hal's sleeve, saying, "Sorry about your mom and grandmother, Hal. It's O.K. I also enjoy being waited upon and I'd like six chicken nuggets and a Coke."

Hal was back with their orders in a few moments, set the food off on the table, sat down.

"You know, Maria, it's good to see you. Have you put on height since graduation?"

"A couple of inches. But I've also slimmed down and that may make me look taller," she laughed.

Hal continued, "You know, I've not seen a pleasant face from our class for three years. In fact, the only person I've seen at all is Sal Lugano, and he's been sort of a thorn in my side, ringleader of a group who don't like me because they think my family threatened them. You know the picture."

      "Yeah, there's some of the same junk in my own family, especially the Quebecois who hate migrants, say they deserted Quebec at a time of need."

      "Don't know anything about that, Maria. Maybe you can educate me 'cause I see only an occasional reference to such hard feelings in the GAZETTE."

      "O.K., in due time, Hal. Right now I've gotta watch the clock 'cause I'm due at the nursing home at 6:45." And she began to eat.

      Hal took a bite of his cheeseburger, began asking her the same questions he'd asked Ed Taylor. To each question she answered directly, occasionally saying "I don't know."

      Hal liked such honesty. She said she didn't mind the smells. "After all, nursing is full of foul human smells, so fish shouldn't bother me!"

      As she spoke, Hal noted her even teeth, her brown eyes, modest make up, her easy smile. He also liked the way she wrinkled her nose and said to himself, "I am glad I didn't study her face when we were in school. It seems fresher this way."

      He was nervous asking about her strength, blurted, "I don't want to challenge your strength, Maria, but do you honesty think you can wrestle traps, throw them on and off the truck, juggle them on the boat, lift the anchor occasionally, all without hurting yourself?"

      "Hal, to be honest, I don't know. But at work I roll 250-pound men and women onto bedpans, leverage them into wheel chairs. In fact," she smiled, "want to feel my muscles," and she rolled up her sleeves!

Hal, still shy, laughed and replied, "No, Maria, I believe you!" They joined in laughter.

When it reached 6:30, Maria said, "We gotta wind this up so I can get to work, Hal. Got any more questions?"

"Yeah, one or two. The easier one: would you like to see my boat and try fishing with me before you say 'Yes' or 'No'"?

Maria answered directly, "Love to! When?"

"What's best for you?"

"Well, I'm off on Saturday, what time?"

"How about meeting you at the dock at 6 o'clock. Most of the guys have shoved off by that time; we'll have less difficulty moving gear and getting to the boat." He said to himself, "No point confronting the gang anyway."

"O.K., that's fine. Now what's your last question?"

Hal scowled a bit, and Maria picked up his hesitation. "Don't worry, Hal, I like you and you won't hurt me no matter what you ask. In fact, Hal, I'll bet I know the question."

"You do?"

"Sure. You want to know if I can take it in a male world where I'm likely to be insulted and you taunted for

having a female sternMAN!" She smiled as she stressed
"man."

"How'd you guess that was my question, Maria?"

"Because we've avoided it for an hour!"

"O.K., what's your answer?"

"Hal, when I say I'll do something, I do it. It's my
French-Canadian background. We're a stubborn lot. I've
also got that tough Franco attitude toward hardship. If
your fishing buddies want to give me a hard time and I'm
really enjoying the job, I'll show 'em they can't rattle me.
A tougher question, Hal. Can you take the razzing you'll
get if you hire me?"

"On the boat Saturday I'll tell you what life's been
like the last couple of months. That will answer your
question."

Leaving McDonald's, Maria held out her hand and
Hal took it, holding it a fraction longer than he had
intended, but told her how glad he was to see her. She
smiled and walked to her car while Hal looked on, waved
good-bye, feeling warmer than he'd felt about another
person for as long as he could remember. Driving home
he thought of many things he would like to have said and
began making a mental list of topics he wanted to discuss
with her. And for all the problems her working on the
"Mary Q" might pose, he could hardly wait to get her
answer. He hoped it would be a resounding "Yes!"

Chapter 5

Rendezvous with Maria

The two days between Hal's meeting with Maria and her arrival at the docks early Saturday morning really dragged for Hal. He did routine chores such as build traps, fetch bait and repair gear. He ran over to Barnsville with a cod for Jerry Moulton when he learned that the Barnsville gang hadn't hit the school of fish. He also asked Jerry to tell his nephew that he probably wouldn't make a decision until Monday.

"Somebody else pendin', Hal?"

"Yeah, she's gonna go fishing with me o n Saturday."

"SHE!" Jerry exclaimed.

"Yeah, one of my high school classmates, dynamic lady."

"But, Hal, you just got through one crisis, why do you want another? You know those guys are goin' ta get on your back if you put a female in the back of yer boat."

"Probably, Jerry, but I think she can take it!"

"Hell, Hal, it's _you_ who has to take it."

"Yeah, I know it's risky, but I may be tempted if she wants to do it after seeing the conditions first-hand."

"Borrowin' trouble, son. Tell me, is she a good-looker?"

"Yup!"

"And that's yer reason, thinkin' with yer heart and not yer head?"

"Oh, it's not that, Jerry. Only had one date in my whole life, to the junior prom. A disaster and I swore off women."

"All the more reason why this 'un's attractive?"

"May be so, but I doubt it."

"Well," concluded Jerry, picking up his cod and heading for the wharf cleaning board. "It's yer own life. Good luck."

"But, Jerry, it's not set in concrete...yet. Sure 'n' tell yer nephew I'll be in touch with him Monday at the latest. Next week's the Fourth, gotta make some decisions soon." Hal waved 'so long' to Jerry, jumped into his truck and headed back to Baysville.

He always enjoyed that crooked road. Dangerous but interesting, and somehow it stimulated his thinking. Knowing it was Friday and he had only a few more hours to prepare for Maria's try-out, another notion flashed through his head. Maybe he'd best check out the boat to see if it was spiffed up enough. While he'd washed her down after fishing that morning, he forgot Maria'd probably want to see the cabin, best show it to her anyway. He drove to the wharf and rowed his skiff out to the "Mary Q." As he climbed aboard, he saw things that he guessed might turn off any woman, especially the cluttered deck. So he rewound a couple of coils of rope, tossed some pieces of a broken trap into the skiff, put away an abandoned roll of paper toweling that had skidded under the gunnels. As he opened the cabin door and stepped down to the lower deck, he was glad he'd thought to double check his "house-keeping." The place was a mess, fish poles and reels thrown onto the cushions on either side of the center walkway which he'd pulled up when he fixed the shaft in May. He began putting the place in order, swabbing the replaced floor boards, stacking the fishing poles back into their overhead racks, returning tools to their places and removing pillows from the front closet. To do the latter, he had to stand on the head, situated in a rather open space at the front of the cabin.

This raised another whole set of questions. "Golly," he queried, "would she have the privacy she wanted?" Though there was a swinging door behind which she could use the head, anybody sitting on the toilet could see or be seen over the door. Too, the cabin door was hanging rather loosely on its hinges. Anybody standing at the wheel could easily observe persons using the head. So he decided to put a latch on the john doors while repairing the cabin ones. Also he noted that the blanket he once carried in the cabin was gone, and he remembered taking it home to wash. So he began a list of things he wanted to bring from home to the boat, including his tiny methane stove on which he made coffee.

Driving home, he reminded himself that he'd have to control his own impulse to urinate over the gunnels any time and any place he wanted. Also, he thanked his "lucky stars" that he'd made the special trip to the "Mary Q." "What a sloppy boat I've been keeping!" he exclaimed to himself as he drove into the yard. He also made a mental note to phone "the ivy college man," tell him that he was unsuitable for the job. He didn't relish that chore but was certain the two of them wouldn't hit it off. It seemed fairly certain that either Ed Taylor or Maria would work out fine.

Looking at the calendar where he'd tallied the number of traps he'd made or 'rehabbed' since his father and grandfather died, he noted that he was up to 279 and had only another week or so to reach his goal, "300 by the Fourth." Studying the calendar also reminded him that he had only two weeks to finish his drawings for the exhibition, so he decided to spend the evening redrawing some of them on the specified art paper. Also, he thought, "Maria might want to see them tomorrow since she seemed interested in what I'd done since school." So in a sense he had a double incentive to work. He became so involved that he failed to hear the phone the first time but finally answered it on the fourth ring.

Hal recognized Maria's voice at once.

"Oh, hi, Maria. I thought you were working."

"I'm on a coffee break, decided to call to make sure you still want me to go fishing with you tomorrow."

"Sure do, Maria, looking forward to it."

"No second thoughts or reservations?"

"Nope! 'Twill be fun showing you the ropes."

"O.K., I'll be at the docks at 6 if my little Omni starts."

"Is it giving you trouble?"

"Not really, Hal, but you know cars. They're great when they're running right, a pain when you need them most."

"Tell you what, Maria. I won't leave for the harbor until quarter to six. If you can't start your car, phone me and I'll come after you if you tell me where you live."

"O.K., that's a deal. See you in the morning."

As he hung up, Hal was joyful. "She's so natural and easy to talk to," he noted. "She seemed pleased I'd drive to Statesville after her." Getting back to his drawings, scattered over the dining room table, he worked until bedtime, poured himself a glass of milk, then retired, wondering if he'd be able to sleep so keen was his anticipation and anxiety about the next day's journey into several unknowns.

Questions flowed through his mind like a cataract: "How will Maria respond to the constant motion of the boat? Will she get upset by the normal clutter of seaweed, ropes, gull's bombarding from overhead? And what about profanity over the CB? Will my Baysville friends and enemies try to embarrass her and me? Will the cabin and toilet facilities be O.K.?"

Then for what "seemed like hours" he reviewed Jerry Moulton's question about thinking with his heart instead of his head. "Maybe I've stayed too much by myself since the folks died? Been lonely without knowing or admitting it? What if we do fall in love? What then? Or say I love her but she doesn't love me? Did I deceive her when I mentioned Jerry Moulton's reservations? And if she thinks I did, will she distrust me?"

As his grandfather clock struck 11 and he checked it out against his red-eyed monster digital, he began to be anxious about not going to sleep. This led him off on another track. "We didn't discuss religion. Not that I'm very religious or go to church, but what if she is an ardent Catholic? No problem now since we can't legally fish on July and August Sundays anyway, but what if she stays on and we fish Sundays after Labor Day? and she insists upon going to church? Well, I suppose she could always go to Saturday night mass? Nor did we discuss days off or money, other things to thrash out if she's interested in the job."

By midnight Hal was so wrought up he decided to get up and take a walk. He threw on an old jacket, went into the backyard where he sat for a half hour star-gazing, noting especially how clear Orion and the Big Dipper were. "Good day tomorrow, lucky us!" he exclaimed. He loved the smell of the marsh behind his house, even at low tide, took several deep breaths of fresh air before the wind changed directions and a warm blast surged from the west, bringing a squadron of mosquitoes. "Damned pesky things!" he muttered, swatting at them futilely and beating a hasty retreat to the house.

When he got back to bed, he decided to concentrate upon the images of his sketches and forget tomorrow. The next thing he knew his alarm was going off at 5 o'clock.

As promised he waited until 5:45 before leaving the house. He was glad that it was fairly warm, maybe sixty-five degrees and NOAA radio promising sunshine and smooth seas, with winds at no more than three knots before noon. As he rolled down the road, he noted an outside light still on at Warren Reed's house and wondered if Warren's son-in-law was O.K.? When he reached the dock, he saw that most of his "buddies" had taken off, leaving the wharf clear. Within a few moments, Maria rolled in with her Dodge Omni, parked and came running to the dock, wearing a sweater and jeans and carrying a plaid woolen jacket in one hand, yellow kid gloves and an airplane bag in the other. "Hi, Hal, made it on time."

"Hi, Maria, right on the button. I see you know it can be cold out there, even in June, bringing warm clothing. We forgot to talk about that. But NOAA promises us a beautiful day."

"Yeah, and I made us some sandwiches and cake. Brought a couple of Cokes. I assume we can stop fishing long enough to have a picnic?"

"Of course." Hal loved her ready smile and added, "You know, Maria, I've been so preoccupied with your coming today that I plumb forgot to bring my usual peanut butter sandwiches and cookies. I'm glad you remembered."

Maria looked up to take in his six-foot height, smiled again and touched him lightly on the arm, saying, "I'm here to learn. Oh! and I brought gloves, thought the ropes might be tough on my hands."

"Super, 'cause I forgot to suggest it. And I guess I have a new role in life, to be a teacher. Remember my fights with teachers over drawing?" They both laughed heartily.

Hal then showed her how to roll the baitbarrel out to the winch where they could drop it into the boat without having to lift much. They then climbed down the dock's one ladder to Hal's skiff, Hal asking, "Want to row?"

"Sure, I've only done a little before, but I'll try." And while she was a bit nervous sliding the oars into the oarlocks, she soon straightened them out, developed a rhythm and rowed smoothly to the "Mary Q." Hal said nothing, simply watched her with admiration and fairly pinched himself that all of this was happening. He was determined not to criticize her, no matter what she did. He'd had enough teacher and family judgment laid on him in his own life and wanted no more of it.

When they reached the boat, Maria remarked. "Hal, I love the name of your boat. Where'd it come from?"

"This was my father's boat, and my mother's name was Mary Quimby. So I decided to leave it in her honor."

"That's nice. I hope somebody does something as nice as that for me someday."

"They prob'bly will, Maria."

At the "Mary Q," Maria seized the skiff's rope and scampered quickly aboard, steadying the small boat against the side. Hal noted her agility, but he did not expect what came next. She held out her hand to help him climb aboard.

"Thanks, Maria, nobody's done that for me since Gramps helped me over the gunnels when I was a boy."

Maria tipped her head back to laugh and her cap fell off onto the deck; they both stooped to pick it up, bumped heads, enjoyed the "accident." "Guess any of us can use all the help we can get. But tell me if I'm too forward, Hal. I confess that I'm on edge about this trip. My whole life may depend upon it. I told you I don't like nursing. Maybe I can be good at this?"

"You'll be good at it, Maria, 'cause your attitude's the most important thing to me. It's so refreshing after listening to some of these men complain, complain, constantly complain!"

"Do you mean bitch, bitch, bitch?"

Hal laughed. "That's prob'bly more accurate as you'll see when we hear them chatter, cackle and crow on the CB."

Hal then showed Maria where he hid the key to the cabin door and boat engine. He also took her into the cabin, explained how the head worked. "And you'll get privacy only if you close the cabin doors and latch these," he instructed, showing her how it was done.

"Nothing like getting down to basics, Hal. Thanks for the tip 'cause I'll admit I was a little nervous about whether or not you had a toilet er- a - head on the boat."

"No running fresh water, but I keep a gallon jug of it and a bucket of salt water on deck."

Hal then made a pot of coffee on the little stove, showed her how simple that was. Then they returned to the deck and Hal started up the motor, remarking, "Usually, I'll do this, but it's important you know how in case I fall overboard or we have an emergency." He also demonstrated the use of life preservers.

"Hey, don't jinx me so soon, Hal. Please stay _in_ the boat!"

"Yeah, but one thing about fishing, Maria. You always have to be prepared for the worst. Look what happened to dad and Gramp..." His voice trailed off...

"What happened, Hal?" Maria asked, looking earnestly into his face, momentarily intense.

"I didn't tell you before because I didn't want to frighten you, but I don't want to deceive you either."

"So what happened, Hal? Please tell me," Maria urged.

Hal then related the story of his family and its traditions, how they were killed by the rogue wave, and how he happened to inherit his father's boat.

"I'm glad you told me, Hal, and I'm truly sorry about them. I met your father at graduation, remember. But you know I've lived all my life in these parts. I've read about these disasters and may have seen the headlines on your father and grandfather but somehow didn't connect them with you. But I guess danger goes with ev'ry territory. Let's push on." Maria paused, then added, "Maybe you'll show me their pictures?"

They wheeled into the dock; Hal lowered one barrel of bait, showed Maria how to lower the other. And they were off for Jocks Rocks.

At first Hal was a bit bashful in telling Maria where they were going. She smiled when she first heard the name of his fishing grounds but noted his hesitancy to say it, remarking, "Look, Hal, you gotta remember that I've heard every kind of profanity and raunchy joke in the book. Nursing classes were full of them, feminist groups, too. Don't let my being a woman cramp your manner of speaking. Say whatever you please. I like the name, 'Jocks Rocks,' for the double meaning and it would be my guess that only men with balls would dare fish around them if they're as dangerous as you say."

Again, Hal could hardly believe his ears. He hated pretension and Maria seemed a model of frankness.

And so the morning passed. Hal showed Maria each step of the lobstering business. He reminded her, however, that she could not touch the lobsters unless she got a license; otherwise he'd run the risk of being fined by the warden. So for three hours she prepped bait boxes, juggled traps, disentangled ropes. At midmorning they stopped to snack. Since there was little or no wind, Hal took the "Mary Q" off shore, a half mile or so from his traps, and let the boat drift. He explained the importance of such a move so they wouldn't drift into ledges just beneath the surface at low tide. As he steered the boat toward the horizon, he felt comfortable about her standing beside him, asking questions about the boat's gauges, rpm, amperage, sonar, etc. He answered both fully and joyfully.

When they'd reached "the place of drift," as Hal called it, Maria rinsed her hands in salt water, opened her picnic bag, layed a clean towel over a box, placed the sandwiches and the cans of Coke side by side. They then sat on the gunnels, teetering and talking as the boat rocked gently from side to side.

"Hal, this is so much fun, and I can't tell you how important it is for me to be here. You're a good fisherman and teacher. Has anybody ever told you that you're gentle?"

"Not really, Maria, they usually call me names like 'Passive' or 'Wimp' or 'Arty' because I've refused to fight most of my life. My drawing and poetry haven't helped. Lobster fishermen aren't supposed to do such things. Even my own family never very much approved. Gentle doesn't 'take' with male machos!"

"And who sez?"

Hal looked down into her salt-licked face and felt the urge to kiss her for her kind comments, but something told him not to. Rather, he said, "_They_ do," and he nodded in the direction of other Baysville fishermen, continuing, "male machos do. And a lot of it came from my dad and Gramps. I loved them, Maria, but they were severe in their male expectations. If I didn't shape up by their standards, I was wrong, very wrong. Even they threatened to throw me off their boats, kept saying just what you heard our teachers tell me, 'Pay attention! Pay attention!'"

"Yeah, I know that scene. My father's the same way. If he knew I was out here with you, he'd probably drown me immediately."

"Maria, do you mind telling me about your family since I've told you a little about mine?"

"Sure. Second generation. My father works as a foreman in the Statesville box factory, supervises five men there as well as one wife and two daughters at home!"

Hal liked the way she stated it, joined her in a smile.

She continued, "It's O.K. for us to smile out here in all this beauty, but it's hard to take when he keeps treating me like a five-year old and my younger sister as three! Mom takes a lot of crap from him, including physical stuff, but she swallows hard and serves as a good Franco wife. He's never been very happy about my independent ways. I've stayed home because it's been less expensive with tuition to pay and housing costs so high in Statesville. But I don't know how much longer I can take his domination. It's sometimes physical, too. He even threatens me when I want to stay home from church, a church I don't believe in. He raises that old priestly scare tactic by telling me I'll 'fry in hell' if I don't go to church to pray and confess. But what good does it do to confess to some old man in a booth, a man I sometimes can't see? I'd much rather be out here with you, listening to your life story and telling you mine. The seagulls would recognize sin more quickly than those old priests; best of all the gulls wouldn't go around talking about their parishioners. Well, you see..."

Since she'd stopped in the middle of a sentence, Hal looked at her quizzically.

"Oh, I didn't mean to get up on my soapbox, but it's all part of my family situation. You can see for yourself where I'm at. I'll tell you more if we decide to work together."

"Thanks for being so frank, Maria. My family was always tight about telling anybody anything, Yankees clamming up."

Again, Maria laid her hand on Hal's arm. "O.K., Hal, we've had our picnic and true confessions, let's go!"

Hal stepped back to the wheel, looked back to see Maria sitting gracefully on the starboard gunnel and staring at Baysville boats on the horizon. She turned to see him staring at her, remarked, "Aren't they beautiful, Hal, backlit by that shining golden water?"

"Today, yes, Maria."

"You seem to be angry with some of them. Can you tell me about it?"

"When we get back to the house." There, he'd told her his intentions.

"Back to the house, Hal?"

"Have to eat, don't we? I've some fresh fish we can fry and vegetables I got coming from Barnsville yesterday. O.K.?

"That would be lovely, Hal, but does that come with today's lesson?"

"Of course, if you'd like."

"I'd like."

"O.K., we've about fifty more traps to tend and we're almost down to the eastern string. Incidentally, I've turned on the sonar because the bottom's tricky on this side. I wanted you to listen to the CB so you could catch some of that local macho color, but it might be distracting. Maybe we can listen going back to harbor or next time."

Standing at the rear of the cockpit, Maria heard the words "next time" and glowed within even to be able to think, "next time." But she said nothing, tilted her head back to let the salt air flow into her nostrils and through her hair. She'd not been so happy in a long time. This was so much better than smiling to please her boss or family.

They tended the eastern string of traps quite routinely. Both boat and crew performed flawlessly except for Maria's getting her foot into a loop as a trap fell overboard. Hal saw what was happening, quickly snatched the rope and barked, "Lift your foot, Maria! Lift your foot!" Maria responded alertly, the trap sank into the depths, and Hal released the rope before returning to the wheel.

"Sorry to speak so sharply, Maria, but I didn't want to fish you out of the water!"

Maria turned toward the rear of the boat. A tear oozed from her eye. She blew her nose, then turned back to Hal. "You did the right thing, Hal, I was stupid to be so careless with my feet. Maybe I..."

"Please don't say it, Maria, you're doing fine. While I was growing up, Gramps fished me out twice! Sorry it happened on your first trip."

On the way back to the docks, Hal remeasured all the lobsters they'd caught, some of which were the largest he'd seen all season. Maria fired a stream of questions at him, about the life of a lobster. Hal told her as much as he knew. But several times he said, "I don't really know, and I don't think the researchers know either. Still lots of study going on, both at the university and with other funds. It's a large commercial business and the State of Maine wants to keep it that way."

Following the compass course Hal had set, Maria steered the boat shoreward while he remeasured the lobsters. Close to the harbor entrance, they stood together at the wheel.

"Want to turn on the CB and listen to some of that macho stuff, Maria?"

"Let's not, who needs it? Maybe next time?"

This time Hal read her "next time" joyfully and casually inquired, "Does this mean that you've decided to take the job even though we've not discussed money and days off? You must have other questions?"

Maria turned to watch the white water at the edges of the trench behind them, gulls following closely, then turned to look up into Hal's eyes. "You know, Hal, I ought to ask such questions, but somehow I trust you enough to know you'll pay me what you can and be honest with me about time off. I think I've made up my mind to join you, but I'll hear what you say about conditions after both of us have slept on it. Is tomorrow soon enough?"

Hal was ecstatic, put his arm around her shoulders and sort of tapped on them. She moved a little closer to him as though to say, "All's well."

Returning to the wharf, Hal and Maria were not quite as fortunate as they'd been when they left. Two of Hal's enemies were pulled up to the dock, crowding the ladder where Association fishermen climbed to street level.

As Hal rowed the skiff to the dock, he said to Maria. "I'll explain later, but I may introduce you as my cousin. O.K.?"

"O.K., Hal.

So when Digger Danes and Walt Carmen turned to glower at Hal and Maria, Hal kept rowing toward the ladder wedged behind Walt's boat. "Gonna be long, Walt?" Hal asked. "Or can we climb across your boat?"

"None of your dang business how long I'm gonna be. And yer can't walk across my boat with yer girl friend, Friday."

Digger yelled, "Me neither. Just wait yer turn."

Hal yelled, "O.K, there's more than one way to skin a cat." Turning the boat around, he rowed around both boats to pull up to a float belonging to Jack Small. He knew Jack would understand without a snarl.

Maria said nothing but stared at Digger and Walt as Hal's skiff passed the two big boats. When the two men muttered, "What yer starin' at?" Maria said nothing.

Back at their cars, Maria whispered to Hal, "Is this part of what you're going to tell me about the Baysville climate?"

"Yes, I'm sorry to say."

"Please don't worry about me. I've been shouted at before. But you'll give me the complete picture?"

"Sure will, during lunch."

Maria followed Hal's truck; meanwhile, Hal was saying to himself, "I'm sure as hell glad to get off that wharf today."

They wheeled into the yard, and Hal escorted her into the house, remarking, "Maria, this place is a mess 'cause I've been sorting pictures and never keep up with the housework."

"No problem, Hal. I love to cook, but I'm not much of a housekeeper. Mind if I walk around and look at your pictures?"

"Make yourself at home while I get lunch."

"May I help, too?"

"In a moment you can set the table and make tea."

So Hal busied himself with fixing the meal, occasionally answering a question which she brought to the kitchen door. Finally he handed her some plates and utensils, and she made a place for them midst his drawings on the dining room table.

"You know, Maria, you're the first person to eat here since dad died. It's been lonely. Wouldn't you know we'd have to 'break bread' in an aroma of lobsterbait."

She laughed, "Can't smell a thing!"

"Lucky us!"

For the next two hours Hal told her the story of being harassed, how his family sort of took over the Jocks Rocks territory years ago and why the "local machos" tried to force him out. He left out the parts about guns and rope slashing.

"But, Hal," Maria inquired toward the end of their discussion, "Isn't there any way you can establish the territory through the law or Association and not have to subject yourself to such harassment?"

"Never been done, probably never will. Ledges under water belong to State and Federal governments."

"The old story of the survival of the fittest?"

"Yup, that simple."

"And you don't enjoy fighting?"

"Nope. That's why I recently came close to giving up fishing and might have if it hadn't been for Aunt Sarah." He then related the story of her pressure.

"Tradition's pretty strong in your family?"

"Goes back almost three centuries, hard to break."

"You're torn between drawing and lobstering?"

"Was," Maria, "until I decided to get some help to fish more traps and make drawing a sideline." With that comment he gestured toward the pictures at the other end of the table.

"And you're going to show me your pictures and tell me which ones you're entering in the contest?"

"Want to see them now?"

"I'd love to, Hal, but I really must get home. I told mom I'd be there about five and help her shop tonight. She doesn't drive and dad won't help her a lick. Can we make it another time, but before you have to put them into the exhibit?"

"Of course, when do you say?"

"You can't fish tomorrow, how's that for you? I promise to give you a decision and it's not too far to drive. Would that be O.K.?"

"Terrific, tomorrow afternoon? But, Maria, don't you want to discuss pay and hours 'cause I really do want you to come aboard."

"Thanks for wanting me. We can work out the details, too, but I want to make a decision independent of time and money. As I said in the boat, I trust you to do the right thing."

As he walked her to her car, she barely touched his elbow, turned to face him, said, "Hal, thank you for a beautiful day. It's meant a lot to me." They shook hands and she jumped in, started the engine and opened the window. "See you at one tomorrow. I'll bring some sandwiches and cookies."

"Great!"

As he watched her drive down the road, Hal felt on Cloud 39.

Chapter 6

Sunday Interlude

At one o'clock sharp, Maria wheeled her Omni into Hal's yard. It was a sparkling clear day. The morning fogs had burned off but still hugged the horizon, a light brown mist. Hal was secretly glad they couldn't fish...with that ominous fogbank hanging there. In fact, he was watching it as Maria drove in.

"What's new?" she asked, tossing her head back as he bent to the open window, then dove into the car to retrieve a picnic basket.

"Hi, Maria. Just watching the fog roll along the horizon and glad we're landbound today," Hal said, as he shook her hand and took the basket.

"Want to eat outside, Maria? Mosquitoes could be bad; you know, one mosquito bites and the world begins to scratch!"

"Never heard that before, but that's a good way to put it. Sure, I can use a little sun to go with what I got yesterday, diving back into the car to get a floppy hat. "Gotta keep the sun off my curly locks," she joked.

Hal was struck again by her simple beauty, and it was the first time since high school that he'd seen her in blouse and skirt. He risked a compliment, "Love the colors of your outfit."

"Thanks, Hal. Feels good to get into something other than that white uniform or jeans."

"Hope you brought plenty of sandwiches, I'm starvin'!"

"Me too, slept late. Only coffee for breakfast."

As they sat in the sun and ate off TV trays Hal brought from the kitchen, their conversation moved easily from one topic to another.

Hal got right to the issue. "Made a decision, Maria?"

"Yes, I definitely want to work with you. Need a change to my first love, the out of doors."

"A decision without reference to money?"

"Yes, but why don't we clear that up now. How's it look to you? I've heard the word 'money' so many times in my life, I almost hate to hear it again though I've gotta be realistic, with car payments and all. Probably I should move outa the house, earn a lot of money, and be on my own. But as I told you, I've resisted. My mother now needs moral support more than ever."

"O.K., down to brass tacks, this is what I can do: I can break even if we catch 50 to 75 lobsters a trip. So how would it be if I pay you $5 dollars an hour for all the work you do while we catch 100 lobsters? Then, for the time being we'll split 75-25 for all we catch over 100, assuming the prices hold fairly steady If things go well, we can be more generous with one another."

"That seems fair enough. What's your average catch, fishing as you now do with about 250 traps?"

"'Bout 125 or so every day."

"So if we go up to 300 or more traps and on beyond, the catch should improve and I'd make more money."

"No guarantee but prob'bly. Of course, some days we won't get out or fog may drive us in early. Hard to predict."

"O.K., Hal, I'll take my chances, just as you do. 'Cause after all, you've got your money tied up in boat and gear."

"Luckily, Maria, dad and Gramps did the buying. I inherited the boat and traps, so that takes immediate pressure off. Got no loans and stuff like that."

Hal paused, then went on. "Are you good with figures and bookkeeping?"

"Not too bad."

"Then during off days maybe you can help keep track of maintenance, costs, help build traps and the like?

"Of course."

"Great, so it's a deal. No, wait...what about time off?"

" How many days a week do you usually fish?"

"'Bout five in the summer, unpredictable in the fall and winter. Biggest unknown's stormy weather, especially in September and October. Might not get out for days, especially around...Jocks Rocks." Again, he hesitated in saying, "Jocks Rocks," and she noticed.

She laughed, said, "'Member, I like the name of your territory."

He smiled warmly.

Then she picked up the conversation. "So you're saying, Hal, that aside from Sundays in the summer, we'd probably fish only four or five days a week anyway?"

"Yeah, that's about it, but we might starve in the winter."

"Then no point in planning days off; nature'll take care of it. Maybe I can waitress part time if we get as far as winter and are still speaking!"

"That's up to you, Maria; we can probably adjust."

"Then, let's go with it," and she held her hand across his food tray. They shook, both pleased.

Hal then asked, "Want me to put it in writing?"

Again Maria gently touched his arm, and replied, "Not really. We both have good memories, and I trust you. But promise me one thing. If I don't do well or you want to modify our agreement that you'll be frank. I want no special treatment."

"I promise."

"O.K., great! Now that that's behind us, what're we going to do this afternoon?"

Hal looked over her shoulder at the marsh, held his finger up into the wind and turned to gaze out at the fog bank. She followed his every move, waiting for a response. "Prob'bly we'd best stick close to shore, Maria, 'cause I'm afraid the fog'll be back in a couple of hours. Do you wanta see some of my drawings before I take them off to the exhibit."

"Of course!" Maria exclaimed.

When they got to the dining room where Hal's sketches were piled, Maria suggested, "Hal, why don't you show me the ones you like best, for whatever reason, then go from there?"

"Fair enough." So he walked around the table, selecting both large and small drawings and handing them to her.

From time to time she exclaimed, "Wow! I like this one" or "This one's terrific!" He placed those in a special pile on the buffet, again admiring the grace with which she walked. Only once during their deliberations did she comment, "Hal, I don't think this one'll make it."

Next they reviewed the ones they'd set aside as she asked, "How many can you enter?"

"Up to eight."

"Hmmm, eight. Has anybody else seen them?"

"Nope."

"Do you think you should ask somebody to look?"

"I've thought of taking them to Jerry Moulton and ask him to check the accuracy of details such as the rigging on this old clipper ship." He held up the picture for her to see.

"Want me to go with you or do you prefer to go alone?"

Hal was a bit jolted by the question because he wasn't sure he wanted Jerry to meet Maria at this time, in view of his comments about Hal's hiring a woman as sternman.

Maria noticed Hal's hesitation.   Something wrong?"

"Not really, Maria, but I must be frank in telling you that Jerry doesn't think that I should have a woman as sternman, argues that I've just been through a crisis with the lobster gang and shouldn't lay myself open to more harassment from the macho guys here in Baysville."

"But you want to hire me, no matter what happens?"

"Of course!"

"Then maybe we should go see Jerry anyway. Once he's met me, maybe he'll help fend off the machos?"

Hal reflected a moment, then observed, "That's true, Maria, hadn't thought of it that way. O.K., let's go, we have the whole afternoon ahead of us and Jerry's always on the docks Sundays. Here, do you mind holding the drawings to keep them flat; I'll get a folder for 'em."

Maria helped Hal get them into a portfolio. As they strolled to the yard, Hal ventured, "Let's go in my truck, Maria, you've not had a ride in it. You may have to drive it at times."

"Never driven a truck. Do you think I can?"

"Of course! It has an automatic shift like your Omni, just a little bigger to maneuver. I'll show you on the way home."

Enroute to Barnsville Hal glowed as he told Maria about every house or piece of marsh that had been meaningful to him as he grew up and the way he always got ideas driving along that crooked road.

She frequently exclaimed, "That's keen" or asked more questions about points of interest. Before they knew it, they were driving onto the docks at Barnsville.

Climbing down from the truck, Hal took Maria's elbow as she tried to balance her body and the drawings. She not only said, "Thanks," but added, "Before I met you, Hal, I might have objected to your help. My feminist bias. But you're the most thoughtful person I've met for a long time. My father would never think of helping anybody, especially a woman."

Hal felt a warm glow but said nothing as he spied Jerry and escorted Maria toward the dock where he was sitting, whittling on one of the model dories that he carved for the tourist trade. She quickly noted his wizened face and gnarled hands and wondered how he managed to carve so smoothly.

He saw them before Hal could speak, "Hi, Hal, who've we got here?"

"Jerry, I want you to meet Maria LaRoux, from Statesville. She's the woman I mentioned the other day. Has decided to be my sternman. Sorry to let your nephew down, but I think she's gonna do just fine."

"Sternman or sternwoman, Maria?" He chuckled as he shook hands, then turned to Hal. "Gotta say, Hal, my boy, you sure know how to pick 'em."

Hal turned bright red. Maria watched with amusement. Then Hal changed the subject. "Jerry, we came to see you for two reasons. I wanted you to meet Maria; also, are you willing to take a hard look at the drawings I'm going to enter into the competition?"

"Uv course!"

Hal handed the portfolio to the old fisherman, commenting, "Jerry, please look at the details. Do you see anything wrong with the eight pictures we've picked to enter?"

"We?"

"Yeah, Maria has helped me sort through the bushels of stuff I've drawn since I was a kid. We think these might be best for the show."

Jerry took a sidelong glance at Hal and Maria, smiled to himself, then turned directly to the pictures, handling them carefully and perceptively, only once dropping a sketch, and uttering a quick, "Sorry!" He then focused on the clipper ship and one of a Marblehead cruiser whose designer he knew, commenting slowly and carefully, "Hal, don't think this clipper would make more than twenty knots the way she's rigged. Gotta tip the mains'ls back a hair, you know, maybe to here" as he pointed the tip end of his whittling knife to the drawing.

"The Marblehead's fine, but you'd bettah add a little to the transom or you might have trouble with the shear."

With that, he handed the drawings back to Hal, resuming his whittling.

"Thanks, Jerry. I knew you'd see more than I saw."

As Maria and Hal turned to leave, Jerry addressed Maria, "Did you say your name was LaRoux?"

"Yes."

"French family from Statesville?"

"Yes, Jerry, if I may call you that?"

"Shuah. Your father's name's 'Oliver'?"

"Why, yes...do you know him?"

"Think so, got some boxes at his factory for a special shipment of lobsters to Norfolk. He waited on me."

"Yes?"

"Pleasant man, very correct and clipped in his speech. Couldn't easily forget him. Saw him again one day here in Barnsville..." Jerry stopped, drew a quick breath.

"And?"

"Oh, nuthin', nuthin'... Glad to meet you, Maria. Good luck working with Hal. He's a nice guy, known him since he was born. Be kind to him." He then turned to Hal to say, "Hal, I'll be seeing Ed tonight. I'll tell him you're all fixed for a sternman, sternwoman. Am sure he'll understand."

"Thanks, Jerry. Glad you could meet Maria. And thanks for lookin' at my drawings. Let you know how I make out in the competition."

"Good luck...to you both..." Jerry said, haltingly.

As they started back the crooked road, Maria was the first to speak. "So he has met my father. I wonder what he knows that I don't about the worthless rat?"

Surprised, Hal asked, "Did you say, 'worthless rat'?"

"Can't help it, Hal. Think he's been cheating on mom for years. And I think she knows it, but won't say anything. That's maybe why she's sick half the time."

Hal saw tears beginning to roll down her cheeks and she began digging in her purse for a tissue. She found nothing so he handed her his handkerchief. "Here, ironed this morning."

She took it and wiped her eyes. Hal remained silent for a mile or two, asking himself, "Why does something always happen along this stretch of road?"

In a moment, Maria handed the handkerchief back to Hal, whispering, "Sorry, guess I'm feeling sorry for myself."

Hal responded, "It's O.K." He paused again to let Maria regain her composure, then added, "Please don't worry. I was going to tell you about this crooked road and what it does for me, also show you how to drive the truck. But you may wanta talk about your family concerns? I suppose we all have 'em?"

"Probably, Hal, but I don't want to burden you. We've got lots of work to do together."

"It's up to you, Maria."

Maria paused a few moments, then said, "O.K., I'll tell you a little, but I may not know the whole story myself. Something's wrong, but I have no proof. Much as my father professes to be a good family man, a good Catholic, a responsible factory foreman, I don't like the way I see him touching women's bottoms in my family, aunts, cousins, and so on. Close and spontaneous as Francos may be in some situations, what he does seems unnatural. Who knows, maybe he was visiting a woman or even a prostitute in Barnsville? Maybe that's what Jerry saw? I've tried to get mom talking about my hunches, but she shuts her mouth as tight as a clam, even grits her teeth when I raise the issue. Early one morning I caught her putting ice packs on a black eye after I'd heard shouting in her bedroom at midnight on one of my night's off. Went down to the kitchen and there she was. She wouldn't let me nurse her, sent me back to bed. Father was nowhere in sight. It's very upsetting, Hal. Probably another reason why I should leave Statesville or at least the house. But I can't, not right now." Again, she rubbed her eyes and Hal handed his handkerchief back to her.

"But you know, Hal, this won't affect my work with you. Once I start anything I block out everything else."

Hal smiled reassuringly, reached across the portfolio of pictures between them to touch her gently on the arm. To his surprise she put her hand over his, sending an impulse through his entire body, one he'd never felt before. He left his hand there for a moment...until he reached the next sharp curve where he needed both hands on the wheel.

Maria spoke, "Thanks for the support, Hal. I hope I won't become a nuisance." She looked off toward the dunes, wondering...

Back at Hal's house, Maria asked to use his bathroom, then prepared to part. "It's been a wonderful afternoon, Hal. Oh, one other thing, when do I start?"

"As soon as you can. I'll follow my usual routine for the next few days, build some more traps, still aiming for 300 by the Fourth. What kind of notice do you have to give?"

"I gave my notice on Thursday after we met at McDonald's. I was pretty sure that I wanted the job."

"Wow! Hal exclaimed, "that's great! But weren't you taking a huge risk?

Maria shrugged her shoulders, responding, "Maybe, but I've got enough self-confidence to take risks. Figured

I'd find something else if you didn't want me. And it wasn't until yesterday, on the boat, that I knew it was going to be O.K." With that she stood on tiptoes, pecked Hal on the cheek, and slid into her Omni.

He touched her arm, asked, "May I phone you?"

She was emphatic. "No, please don't phone the house, Hal. Let me call you from the nursing home." She held out her hand again. Hal gripped it tightly, walking along beside the car as she rolled slowly out of the yard.

Chapter 7

Gnawing and Knotting....

Hal was restless the days between Maria's Sunday visit and the Fourth. While tending his traps mornings, he saw her face in the swirling water and engine gauges. Building traps, he found his mind wand'ring and had to concentrate for fear of hammering a thumb or cutting off a finger on the circular saw. She telephoned Tuesday evening, but they had only a brief conversation since she was "on a coffee break" and had to "rush back to duty." He couldn't sleep for thinking of her and began to wonder if he'd been smart hiring her since he seemed to be falling in love. This led to question after question, fantasy after fantasy, making sleep even more difficult to come. He'd never been successful counting sheep in these circumstances, so he began imagining his routine of tending traps, using his hydraulic winch to lift one after the other from the water. And if he drifted back to Maria and how she might complicate his life to fulfill Jerry Moulton's prediction, he quickly resumed locating his traps off Jocks Rocks and going through the routine of hooking a buoy with his boat hook, threading the open pulley, balancing a trap on the gunnels, opening the cover, checking its contents, baiting it, returning it to the bottom, moving to the next trap, etc. Only while eating meals that week did he allow himself the luxury of speculating on what their lives might be...on the boat, on the land.

But finally, Friday morning came; at exactly 5:30 Maria and Hal drove onto the wharf simultaneously. The Baysville Gang was out in full force. As Hal and she walked toward the baithouse, Digger Danes looked up, and sarcastically observed, "So Kid Hal can't make it by himself, needs a Woo- M-A-N to protect him?"

Hal paid no attention, brushed quickly past Digger and his handful of pals, who also stopped preparing for the day's fishing to toss killing looks and mocking words toward Hal and Maria. Together they walked to the baithouse where Maria rolled out both barrels by herself. As Hal did other chores, he whispered, "Don't roll them too close to the edge of the dock; we'll move them closer when we pull the boat in."

But as they carried their gear to the top of the ladder, Walt Carmen shoved Hal against Maria, almost knocking her down. When Walt made a second lunge toward them, Maria grabbed his arm and in a split second used his momentum to toss him over her shoulder into the harbor.

An "Ahhhh" went up from the group, and Maria turned to face the gang, speaking firmly but calmly, "Look, I don't want any trouble with you guys, but I didn't learn judo and karate for nothing. There's more tricks where that one came from. You'd better fish that guy out and let us do our work. We're here to make an honest day's pay just as you are. Leave us alone and we'll leave you alone. There's been enough trouble already since Hal's father and grandfather died. So cut it!"

With that speech she quickly climbed down the ladder. Hal's mouth was ajar.

Jack Small, who had come to Hal's side during the Fishermen's Meeting, also spoke in low but firm tones to the knot of men, "Look, she means business. Better leave them alone. If you up the ante, you'll have me and the law on your backs. Violence around this place has gotta stop sometime. It had better be NOW!"

The men stared daggers at Jack and still grumbled as Hal and Maria rowed their skiff to the "Mary Q."

Hal was the first to speak. "Wow! You sure scared me, Maria. I didn't know you knew judo and karate."

"Had to learn them to defend myself, Hal. Hated to do that, but I thought they'd better learn right off the bat not to mess around with me."

"Can you teach me judo?"

"Of course," Maria laughed, "but are you sure you still want me to be your sternman?"

"More than evah!"

When they got back to the wharf with the "Mary Q," the few men present gave Maria a wide berth as she maneuvered the barrels onto the winch and lowered them to Hal, still in the boat. Only Bill and Ray yelled obscene remarks at her, but she mostly ignored them, turning only once to scowl.

When they reached the open ocean, Hal previewed their day's work. "Not only do we tend traps today, Maria, I'd like for you to learn knotting techniques for rigging them, a combination of bowlines and half hitches. If we have time this afternoon, we'll prep some traps back at the house to set tomorrow. I thought we'd start today by continuing what we were doing last Saturday. O.K.?"

"O.K. So I'll begin the day by fixing bait? Look," she laughed, as she took a spring-type clothespin from her jeans pocket and pretended to put it on her nose. "I came prepared."

That sent them into gales of laughter.

As Hal wheeled the boat into position to pull the first trap, he pinched himself to have somebody aboard with a sense of humor, to lighten his spirits, shed his fear of violence.

The day went well. Lobsters were plentiful, and they caught 138. The only disagreeable part of the day: Hal asked if she wanted to listen to the machos on the CB.

"Of course, may as well get my ears burned. I'm not out here to enjoy ALL the scenery!"

Sure enough the Baysville "brats," were having their fun with Hal and Maria. Some of their comments were pretty raunchy. Although they did not address one another by name, Hal identified each voice for Maria.

Bill remarked, "Must be a whore to work for Hal? Wonder where he met her?"

"Lives in Statesville," Ray said, "father's a foreman at th' box factory."

"They piss me off!" Walt exclaimed. "Got any dirt on her? I'll get her ass for tossing me overboard!"

Maria and Hal gave one another knowing glances as they heard several guffaws. Hal felt embarrassed but continued with his routine. Maria, too.

"Nothin' on her, but that father!" Sal Lugano insinuated.
"Saw her at the Nursing Home one day when I was there visiting my mother-in-law. Wonder why she left?"

"Dunno, but she must be both dumb and crazy to work for Daniels, must wanta stahve."

"Hell she's crazy," Ray interjected. "If they work their asses off with 500 traps, they'll make it tough for us."

Sal guffawed loudly, then said something that incensed Hal. "Hey, you guys, why ain't somebody using the 'F' word today? Think she and Hal may be list'ning? They both know it. Prob'bly use it themselves, especially in polite company," he added in a deep voice, sarcastically.

Jack Small interjected, "Look you guys, why don't you shut up? There's plenty of room out heah for all of us. Leave 'em alone. She's already showed you not to mess around."

At that point Digger spoke up, "Wonder if she's Superwoman an' can stop a speeding bullet?"

Jack screamed over the CB. "Look, Digger, Elmer Kruse and all th' deputies in the county suspect you for Will Bryant's murder last year. And they know you guys got it in for Hal Daniels. If you get into shooting, you're gonna rot in jail 'cause I'll sure as hell testify <u>against</u> you!"

Digger retorted, "Look, you bastard, I could kick your ass, too. You've sounded like a traitor ever since you made fools of us when you stood up for Daniels at our last meetin'. Whose side you on anyway?"

Jack replied, "I'm on...ALL our sides 'cause I don't want any trouble. We don' need it. Our reputation's gettin' worse all over the Bay of Maine. I repeat what I said on the wharf, 'LEAVE 'EM ALONE!'"

As they headed for harbor and Hal shut off the CB, Maria came to the wheel to stand beside him. As he looked down at her under his long black visor, already beginning to acquire the quizzical eyes fishermen develop, he said, "Well, there you have it, the bitter and the better. What d'ya think?"

"We'll weather the storm. Who guarantees certainty?"

By noon they had caged the lobsters, secured the "Mary Q" on the mooring, put their gear away and were back at the house having a snack.

"Int'resting morning, Hal, but it occurs to me that we didn't bargain about your feeding me lunch!"

"Fringe benefit! Besides, I enjoy your being here, and aren't we going to rig traps this afternoon?"

"That's what I bargained for."

"When must you be home? And how's your mother?"

"Hanging in there. I'll be O.K. if I can be home by five to help her with dinner."

"Good!"

So they rigged fourteen new traps, Hal teaching Maria how to secure the heads, run the warp through the heavy oak cage, and coil the ropes for attaching the buoys. Still on CLOUD 39 while working with her, Hal felt awkward in asking, "When can you show me some judo or karate tricks such as you used on Walt this morning?"

"Whenever you wish."

"Before you leave this afternoon?"

"Sure."

About 3:30 they stopped working on the traps; and, finding a shady spot behind the house on that hot July 5th afternoon, Maria put Hal through some judo paces, even daring to throw him to the ground and encouraging him to toss her likewise though he expressed fear of hurting her. After such "falls" they laughed as they held out their hands to pick up the other. "You're learning beautifully," she said.

"How come you didn't tell me you could do this?"

"Didn't think to. Oh, it passed through my mind when you told me the story of their strong-armed tactics, but I forgot it until the emergency this morning. Really, Hal, it was quite spontaneous."

"Maria, tell me, are you afraid they'll start shooting?"

"Are you?"

"Not really 'cause I think they're cowards. But I do have guns aboard."

"Yeah, I saw them when I used the head this morning. I don't know how to shoot a gun and don't particularly want to, but maybe you can teach me? If they know that I can, maybe they won't start?"

"O.K. it's a deal. You show me judo and karate and I'll show you what I've learned about guns. By practicing here in full view, they may get the message and leave us alone."

"Look, Hal, we've gotta get one thing straight. I understand why you may feel you have to treat me with kid gloves and protect me. I admire that. I can imagine how the women in your family may have been brought up and how they raised you. It was a different generation. But, Hal, we've got work to do now. You hired me to do a job. We have to respect that. We both know fear can be a good shield. But we also know that there's no such thing as certainty and we'd be wise to go with that, Hal...." Her voice trailed off as she looked across the marsh where the tide was nearly full.

Hal had not expected such vehemence but he embraced her aura and remained silent, following her eyes to the tide.

Finally, she turned back to him, remarking, "Gotta go, Hal, if I'm gonna get home by 5. See you in the morning." She touched his sleeve, looked straight into his warm brown eyes and said, "Thanks for everything." She was gone in a moment, leaving Hal standing in the yard and facing another lonely supper and busy evening putting his drawings into final shape.

As he concentrated upon his drawing after supper, he mused about the ways in which his life might change with Maria coming aboard. He also puzzled over the differences between enjoying solitude and feeling the pangs of loneliness. He even pulled out a sheaf of poems, read them in an effort to see which feeling had been most important in writing which poems, sorting them into two piles. Looking out into the stars later in the evening, he had an intense sense of being very puny.

Chapter 8

At the Edges of Crisis

July passed quickly. Hal and Maria soon found the rhythm of tending traps, preparing and setting new ones; by working diligently every afternoon they had set 400 by July 27th. They practiced the martial arts and developed further skills shooting both pistol and rifle. She applied for her lobstering license and announced her intention to attend the early August Fishermen's Association meeting. She also accompanied Hal when he entered his drawings in the Statesville competition, and they occasionally ate a late afternoon snack at McDonald's following delivery trips of lobsters to the Portland market if local retailers wouldn't buy them.

On one of these occasions, the last day of the month when Maria's car was in a Statesville garage near McDonald's, Hal asked Maria if she would like for him to meet her family when he took her home. Maria immediately responded, "No, Hal! My parents would misunderstand."

But as she stepped from Hal's pick-up, her father drove into the yard of their one and a half story, shingled cottage and saw Hal. Maria quickly re-opened the door and said, "Hal, leave quickly, please, I'll meet you in an hour at McDonald's."

So, Hal, sensing something wrong, put the truck into gear and took off quickly, continuing to wonder about this shadow figure, Oliver LaRoux?

Having an hour he drove around his old high school grounds and on into the back country behind

Statesville, enjoying the golden sunlight on the fields of corn, beans and squash, regretting he didn't take the time to do it more often.   But he nervously eyed his watch so he could get back in time to meet Maria.

Entering McDonald's, he grabbed a table. He wasn't too hungry because they'd eaten in a Portland suburb. Moments, minutes, a half hour passed as Hal worried. Finally Maria showed up, dishevelled and breathless. She spied Hal and immediately joined him. He greeted her warmly but properly.

"Want something to eat before we talk?"

"No, just a cup of coffee, Hal. Let me take a deep breath before I tell you what happened."

Hal got the coffee, returned to set the cups down, then slid awkwardly into the seat opposite Maria and stared into her soft brown eyes. He could see she'd been crying. But he kept a discreet silence.

Wiping her eyes she began, "Hal, what just happened shouldn't happen to a dog. Father saw your truck, saw me get out, and was waiting for me at the door, his belt in hand. As I started toward the kitchen to help mom with dinner, he grabbed me by the shoulder and screamed, "'Where've you been? Who've you been with?'"

I told him that I worked for you, that we lobstered together, that we'd been to Portland to market lobsters. He went into a tirade, called me 'a liar and good-for-nothing headstrong deceiving bitch,' and he forbade me to work for you any longer. I shouted back, told him I was twenty-one and would do as I damn well pleased.

Mother, having heard the shouting, came to my side, and he yelled at her, 'Did you know about this?'"

She said, "'Yes, it's a good job and she's doing well.'"

"'And you didn't tell me?'"

"'You've gulped your dinner and been out every night for months. How could we talk about it?' And all of this, in French, of course."

"With her last remark, he screamed, 'And you're a bitch, too. You've both deceived me. I want no daughter of mine working for some Yankee son of a bitch. You'll work with our kind or you won't work.' Then he slammed mom into a corner where she slumped like a crumpled rag."

"I rushed to comfort her, but father was right behind me, tried to pull me away, tore my blouse half off as you can see," and she turned her right shoulder toward Hal where he could see she'd wrapped the rip tight to disguise it. "Then he dug his fingernails into my shoulders and struck my back several times with the buckle end of the belt. I tried to stand up and give him a couple of karate chops, but he held me down with his knee." Again Maria turned slightly where Hal could see the red welts and blood stains on her thin blouse.

Hal felt sick to his stomach as he listened, reached across the table to cover her hand with his, saying, "Maria, I'm sorry and feel ashamed that I'm part of the problem."

"No, Hal, please don't feel guilty about any of this. As I've hinted a couple of times, I was almost sure that this family crisis was coming. But I've told you only a little. I hope this doesn't upset your life or your belief in me."

"It won't," Hal assured her, holding her hand more firmly and asking questions in rapid order. "But, Maria, what will you do now? Do you think your father followed you here? Will he want to beat me up, too?"

"I doubt it. After he finished whipping me and I'd nearly collapsed, he swore a long oath about 'Yankee bastards' like you, rushed out the door, threw himself and his coat into his Ford, and gunned it down the street toward Barnsville. I stayed to help mother change and get into bed, made her a little soup, saw she was comfortable, washed a little, got my car from the garage and came directly here."

"You poor dear."

"Thanks for being here, Hal."

"What're you goin' to do?"

"For Mom's sake I've gotta go back to that hellish house. And I do mean hellish. I've told you only part of the story. Father tried to rape me when I was in high school so I kept a knife by my bedside. He knew it. Also that I was learning judo and karate. But for all of that he tried again to rape me as recently as April, caught me outside the nursing home, forced me into his car. I fought him, but he delivered me a bloody mess right in front of the house. Maybe you saw the scratches on my neck when we first met in June?" And she turned her head so Hal could see them. He reached across and rubbed the back of his fingers gently across the scars and wondered to himself why he'd not noticed. Maria continued, "It was then that I decided I had to make some changes. It's true, he didn't know I was working with you."

Maria took another deep breath and continued, "No, mom lied for me, said I was waitressing days and at odd times and that's why I was around the house at night. Didn't see him much anyway 'cause he was out helling around somewhere. I think that your friend, Jerry Moulton, knew this when we talked with him that Sunday. I'd must ask him some more questions."

"Sure. But what're you goin' to do <u>now</u>, Maria?"

"I gotta home tonight, Hal, to see if mom's O.K. He probably won't come in 'til long after she's asleep. They sleep in different bedrooms."

"Do you think your younger sister knows?"

"Kathie's away to summer school taking a pre-college advanced placement course. But she must suspect 'cause he once tried to attack her. I also have a hunch that he abused her sexually when she was a baby. All this stuff, you know, makes for what somebody called 'the great family lie.' Nobody wants to talk about it, not even children. Even Kathie and I haven't discussed it. It's like a taboo topic. And you know from the TV news that children aren't believed or taken seriously in court even when they do talk."

"I know, but never realized it was so serious. You know how isolated I've been all my life. Is there anything I can do?"

"No, Hal, this is something I've gotta solve myself. You've been kind to listen. I love working with you and won't be intimidated by father. I just hope he doesn't do anything rash to mom. I've gotta go now. See you in the morning at 5:30."

"O.K., Maria. You've my phone number if you need me. Haven't slept well lately so don't worry about waking me."

They walked out of the restaurant together, and she gave him an ever-so-slight hug before slipping into her Omni and driving off.

As he climbed into his truck, Hal decided to go home via Barnsville. If Jerry Moulton happened to be home, he intended to learn more about Oliver LaRoux.

Hal was not sure if it was fortunate or unfortunate that Jerry wasn't there. As he drove back his favorite crooked road, he mused. "Should I stick my nose into what looks like a 'messy family affair?' How long can Maria live in that house without getting hurt badly? Can she really protect her mother? How will all this affect Maria's performance as a sternman? Should I encourage her to move out? Would she by any chance like to move to Baysville?" And, of course, this led to thoughts about her moving in with him. After all, his big old house was almost empty. That in turn led him to recall his family heritage and his parents' frowns whenever they discussed the new sexual morality they saw all about them or on HBO. He could still hear his mother's high-pitched voice, saying, "Why the very i-d-e-a! It's im-moral!"

Slumping into a chair and nearly falling asleep when scanning the local paper, he gasped, "What a day!"

And while he didn't normally drink, he had a beer before going to bed and hoped it would relax him enough to put him to sleep, muttering to himself, "4:45 will come early enough."

He was heavy into a dream about lifting nets and coils of rope onto a shrimp dragger when he heard a bell ringing far, far away. He didn't know how long it had been ringing when he woke enough to realize it was his phone. As he reached for the receiver and looked at his red-eyed clock, he saw that it was 1:17 a.m. To himself, he said, "Maria must be in trouble."

But, no it was Jack Small. "Look, Hal, I hate to call you at this time of night, but I had to get up to piss, looked out the bathroom window and saw a light in the cabin window of your boat. I watched it for five minutes to make sure it wasn't a reflection of the wharf street light as your boat swung on the mooring. I think somebody's aboard, 'cause I saw a shadow cross the porthole a couple of times before the light went out. You may want to row out to see what's happening. I'll meet you at the dock and go with you...if you want?

"No thanks, Jack. Can I borrow your skiff if mine's gone?"

"Sure. Better take your pistol, too, in case any of the boys are into funny business."

"Thanks for calling, Jack. I'll let you know what I find."

On the way to the dock, Hal ran Maria's McDonald's story through his mind, over and over, but hoped it wasn't related to this. "And, wouldn't you know," he thought, "my guns are locked in the cabin....I hope! But," he mused, "only two people in the world know where I hide the key, Maria and Jerry."

Sure enough his skiff was gone, so he slipped noiselessly into Jack's, rowed quietly toward his boat, now totally blacked out. When he approached the "Mary Q," he held the skiff away from the port side of the big boat, to prevent a bump, kicked the little boat away, got aboard silently while looping the rope around the winch. He was tiptoing across the deck when a light appeared in the cabin.

"Hello, who's there?" a voice whispered. It was Maria.

Hal also whispered, "It's me....Hal."

"Oh shit!" came the reply. "I've disturbed you."

"It's all right, Maria, I half way expected it to be you. May I come in?"

"Asking to come into your own boat, Hal? That's a nice twist," and Maria laughed as he found her in light blue pajamas, propped up on pillows, her feet covered with the blanket he now kept in the cabin. Hal could not but notice her shapely curves and the soft shadows on her face and neck, seemingly sculpted by the dim cabin light.

"How'd you know I was here, Hal?"

"Our friend, Jack Small, saw a light in the cabin, phoned me, thinking some of the boys might be playing games with our boat. He offered to come with me, but I thought it might be you. Maria, what's happened and why didn't you phone or stop at the house? You don't have to sleep in a place like this, cozy as it is."

"I heard you say 'our' boat, and believe me, Hal, this cabin is heaven compared with my home. It happened again."

"What happened?" Hal urged.

"Father came home half drunk, found mother asleep, hammered on my door, called me obscene names, said I was impulsive and he'd hold me down and screw me the next time he got me alone, also threatened to kill me if I ever returned to Baysville. Among other things he said, 'I've got AIDS and I can kill you and nobody will ever know it!' With that he staggered down the hall, went into his room, and I soon heard him snoring. I packed some of my clothes in two grocery bags and hurried here! Thanks to you, I had a place to come." And she began to cry, reaching for tissue.

Hal moved across from one bank of cushions that formed a bed on the port side to sit beside Maria, take her right hand between both of his. He was both nervous and awkward. Close as he'd been to Maria during the previous month, he was not sure of the next step. He felt his heart beating faster. And yet for all of this, he gently stroked the back of her hand and forearm. She continued to weep but didn't resist. When she stopped crying, they looked at one another and he fastened upon her glistening eyes and started a poem aloud, "Your eyes are deep springs of tenderness and caring..." Then he stopped abruptly, again unsure of himself.

"That's beautiful, Hal."

Hal began to speak again, haltingly. "Maria, this may not be the best time to tell you what I'm feeling...but you must know that I've been falling in love with you ever since I first saw you at McDonald's...that seems half a lifetime ago. It also seems like yesterday." Hal paused to get his breath as Maria looked at him intensely. He continued, "And the more I've tried to avoid your grace and caring and loyalty as we've worked together, the more I've thought of you, seen your face in almost everything I've looked at...dreamed of you...also couldn't sleep thinking about you. It's almost like what I've read in zen." He took another deep breath, added, "Can you believe me?"

"Of course I believe you, Hal, it's happened to me, despite myself. Hal, I've fallen in love with you, too. You are tender, kind, considerate and caring." With that she lifted against him, put her arms around his shoulders and gave him a strong hug. He could feel her breasts against his chest and it moved him. But she soon fell back with a tiny cry, obviously in pain.

"Maria, what's the matter?"

"It's my shoulder and back, Hal, where father dug into my flesh and whipped me. Guess I've irritated it again. And she leaned forward to lift the back of her pajamas so Hal could see."

"It's inflamed and looks awful, Maria, let me get some ointment." Hal crawled across the cabin to the First Aid kit, found the salve, returned to Maria's side, asking, "Do you want to put it on or shall I? You know how shy I am."

"Don't worry, Hal, I'm just human. You do it. You can reach it better than I." Again, she unbuttoned her pajama top modestly, pulling it up to the back of her neck, and Hal applied the ointment. He saw her grimace as he tried to sooth the flame and took her hand and said, "Squeeze if it hurts."

Squeezing hard a couple of times, she said, "See what I mean, Hal. No matter how tough you try to be with these macho guys, your gentle side always shows. Now I'm glad I looked over your shoulder to see your sketches when we were in school. I'll hug you a little less violently to prove my point." She got up on her knees, put her arms around Hal's neck and kissed him hard on the lips.

Aroused as Hal was, he still felt bashful and awkward. But he responded warmly as they embraced for fully a minute before Maria slumped back against the pillows, smiled and winced, smiled and winced in fleeting rhythms.

When tears welled up in Hal's eyes, Maria handed him some of her damp tissues, saying, "Sorry they're not dry, Hal."

He laughed through his tears, leaned over and kissed her again. She yielded warmly. Through a porthole she saw the waning moon, gourd-like and orange. Pointing across the dimly-lit cabin, she whispered, "I wish that were a full moon, Hal, but I think we can use that one as a good omen. I'm worried only about one thing."

"What's that, Maria?"

"You may feel pity for me."

"No how, Maria, I've watched every move you've made as we've fished, your hard work at your job, your willingness to learn and bear with my timidity and horrible mistakes, your flexibility, your sense of humor. No, dear, those things are important to me. No, I don't pity you. I adore your strength and independence. I'm just angry at myself for not being more realistic at McDonald's last evening. I should have told you that I loved you, and brought you home to avoid what you experienced. Do you think I could have convinced you?"

She heard him call her "dear," but answered, "I dunno, Hal, I dunno. I wanted to help mom 'cause I feared father would become violent to save his own damned face. I also thought he'd blame us, the victims, 'cause he's hidden behind his religion and claim to be a good family man. I've seen a lot of it in my Franco community, Hal. Of course it doesn't apply to all the people of my heritage, but it's the old story of the vital lie to keep the world out of family affairs and away from the skeletons in their closets. I saw it as an adolescent. I've seen it while nursing. A woman would come into the hospital, battered almost beyond recognition. But ask her what happened and she'd say, 'Nothing!'"

"You know, Maria," Hal said, holding her hands in both of his and noting that they were nearly as calloused as his own, "I had a strange, maybe even a <u>dumb</u> question coming from Barnsville after I left you last evening. I'm not sure I know how to ask it. But you know I keep telling you about that road. But how do I ask it?"

"If we're in love, dear...and I heard you call me that, can't we ask whatever we wish anytime, anywhere? Why don't you just say it like you heard it in the pick-up? Somehow I feel you'd never say anything to hurt me."

Hal paused a moment, held her hands against his chest, then blurted, "To me, your father sounds mad. Why would he want to destroy his own flesh and blood? Can it be that you were an adopted child?"

Maria sat bolt upright on the cushions, stared into Hal's dimly lit face, tightened her grip on his hands, paused as though bewildered, then practically screamed, haltingly, "What!   Could it be?"   She looked away, reflected a moment, turned back to Hal, "S'pose that's the reason?"

After giving Hal another hug, she settled back calmly, looked at Hal and said, "Dumb question nothing, Hal. If I was adopted, it might explain more than the attempted rape. I saw my sister being born so I know mother had her. But after that I became a kind of second-class citizen with father.   And yet," she added incoherently, looking off toward the moon, "He tried to rape her, too."

For a long time they just held hands and looked at one another...until she finally broke the silence, "Hal, that may be the most important question anybody ever asked me in my whole life. I'll confront mom. She must have heard father screaming last night. She might talk to save her own face."

Both sat quietly looking at one another, Maria shaking her head in disbelief. Hal finally spoke.

"But, Maria, asking your mother's gonna have to wait 'til morning. Right now, how're we goin' to deal with you, us, and <u>n o w</u>? We may value risk and uncertainty, but the sun's gonna come up tomorrow as sure as we're sitting here in this half-lit, uncomfortable cabin, pledging our love...<u>now</u>. Those guys out there will still be on our case, no matter what we do or don't do. I think you know how much I've respected my distance from you, almost treating it as sacred. And that's my problem, Maria, my rearing, my isolation from women. I didn't want anybody to get the wrong ideas. How can we keep that distance while lobstering, yet grow together? Now I'm less fearful than I was just an hour ago, but tomorrow I'm going to want to kiss you at every lobster pot!"

Maria laughed, kissed Hal again, remarking, "Here's one for the first trap!" She then looked off through the porthole, reflecting. "Hal, I was very impulsive coming here tonight. I hope I've not endangered your life. I'd never do that for the world. But I followed my heart. Also I was stymied and didn't know what else to do. I don't have friends I can trust in Statesville. Too unorthodox to suit them. My month with you has been beautiful although I've lived in two worlds. It's been beautiful even if you turn me out and our contract ends tonight. Nothing can take that month away. I trust you. But I hope I've not set another trap, holed us up in this cabin where the machos out there can treat us like rats." With eyes glazed and looking off toward the moon, she took both of Hal's hands between hers and put them to her lips. In a moment, she asked, "What do <u>you</u> think we should do, Hal?"

Hal also turned to take a long look at the moon, studied his watch a moment, then looked back to Maria as she searched his face for some sort of answer...all the while keeping his hands between hers. "Maria," he began,

"Would you come live with me? You know some of the house. There's plenty of room if we don't let the ghosts of my Puritan ancestors scare us. I'd give you all the time and the space you want..."

Maria studied Hal's earnest face, held his hands the tighter, observed, "I can imagine what the ghosts might say!"

"Yer right. They would frown, but frankly, I don't care. So long as we love each other, who cares what anybody says?"

"I agree and I yield," Maria said, climbing into Hal's lap, one arm around his shoulder while trying to favor her back. "But go on."

"We must be careful for a day or two. We've gotta be prepared if your father shows up. Want nobody to get hurt. So I propose you stay here for the rest of the night. We'll lock the cabin door so nobody can come in, but you can get out if you need to. Keep the lights off. Sleep until I get back. I'll sleep late myself, say 'til 8 or so. By that time the macho gang will be cleared off the docks. NOAA Weather predicts another scorching day tomorrow and the sun went down a red ball. There's water and apples there in the locker. I'll bring your breakfast and lunch. When we're through fishing tomorrow noon, we'll move right back into routine. Let me have your keys, Maria, I'll drive your car home now, put it in my unused garage so your father can't see it if he comes nosing around. I sometimes walk home and leave the truck at the dock anyway, so nobody will be the wiser. What do you think?"

"Sounds O.K. to me. But are you sure you want to take me home to all those family ghosts?" Maria jibed.

"We can always bury them," Hal said, leaning against her, held her tightly by the upper arms to protect her back, and gave her the deepest kiss he knew how. She responded in kind.   He got to his feet, pulled the blanket over her, kissed her again, whispering, "Good night, dear, I'll see you about eight." She nodded, and slid further down under the covers as he locked the cabin door and left quietly.

For a moment at the dock he was frightened when he stumbled over a stray tom cat, claiming his quota of fishbait. But he locked his truck, drove Maria's car home to the empty garage, covered its one window and bolted the door behind him before striding toward the house...musing, "What an amazing day!"

Chapter 9

Sound and Fury

Hal was awakened on that First Day of August by heavy pounding on his porch screen door. It reminded him that he needed to tighten up the screws so it wouldn't rattle when people knocked. As he drew on an old oil slicker coming down the stairs, he said to himself, "What 'n' 'ell," seems I just got to bed. His red-eyed clock told him it was 6:55, just five minutes before the alarm was due to go off. The sun was long up.

As he ambled sleepily across the porch, he saw a strange man standing on the stoop, hammering on the door. He guessed who it might be, but hoped his guess was wrong.

He assessed the man as about 5'10", thin, swarthy, black hair, with a neatly groomed mustache, heavy circles under his eyes, and dressed rather nattily in suit, white shirt and dark blue tie. Before Hal had the inside door fully opened, the man thrust his chin against the screen and blurted, "I'm Oliver LaRoux and I'm madder than hell. What have you done with Maria?"

"Oh hello, Mr. LaRoux, like to come in?"

"No, I'll have no part of you or your hospitality, you gawdamned Yankee no-good fisherman."

Hal maintained his composure, coolly remarking, "Since I don't know you, Mr. LaRoux, and you don't know me, I don't understand what you're saying."

"You DO know what I'm saying. You've stolen my daughter from me on this trumped-up idea that she can become a lobster man. You're a man, she's a woman. And I won't have her screwing around with you and destroying her life, after all the money I've spent on her schooling, becoming a no-good fisherman. It makes me sick to my stomach! If she's here, I want to see her immediately. Her car's not at the wharf. Some of your buddies down there say she works for you and don't think much of it, say they've not seen her car since yesterday. And I know it ain't home, 'cause she stole away in the night."

Hal observed that Mr. LaRoux did have a fair command of English, spoken with a slight French lilt. He was glad the screen was between them. But he decided it best to "play ignorant" for the time being and let the guy talk himself in or out of trouble.

"Well, Mr. LaRoux, you can talk to every man on the dock and you'll get all kinds of ideas about me, but they probably won't tell you anything about fishing."

This touched him off, "Gawdamn you, you thieving bastard, I don't want to know anything about fishing! My people have had it in for you and your kind ever since they arrived from Quebec to slave in the milltowns of Maine. Your kind kept our people down. And you can't deceive me. We don't like you and I know you don't like us, call us names, laugh behind our backs, tell raunchy ethnic jokes about us. And I won't have my Maria slaving for you, and you making her a slut. I won't have it."

Hal continued to be both cool and polite. "Mr. LaRoux, you're entitled to your opinion. I know Francos who have different opinions. Why don't you go home...or to work...or wherever you want to go but leave your daughter and me alone. If I want to pay her well to work for me...and she does a good job...and she wants to work for me, she's twenty-one and can make up her own mind about what she wants to do."

"Is she here now?"

"No!"

"Do you know where she is?"

"Why don't you let <u>her</u> get in touch with you and tell you where she is, Mr. LaRoux? I've never been an informer, and I don't intend to rat on anybody now."

"If I come down here with a gun, I'll shoot you."

Hal maintained his composure. "That's a nasty threat, Mr. LaRoux. If I were you, I wouldn't try it. People in this town would treat you like you've never been treated before...and I'm not kidding. They can call one another names, but they don't like uninvited outsiders to come in and call them names or tell 'em what to do. You may hate old Yankee families like mine, but please, Mr. LaRoux, don't underestimate us. We hang together when we need to."

La Roux moved backward and nearly fell off the granite step. Still hopping mad, he tetered awkwardly, looked around as though pondering what to do next. Hal stood his ground on the inside of the screen door, hoping LaRoux wouldn't make any sudden moves.

The angry visitor tried another ploy, "When do you intend to go to the wharf today?"

"When I get around to it."

"Then I'll smell out the place for myself."

"Go ahead. The road's open 'tween here and there. But when you get to the docks, remember, that's private property. Better keep off as the sign warns. The constable knows what to do with violators."

"Your good buddies didn't tell me that."

"Maybe you didn't threaten 'em."

LaRoux paused, finally walking slowly back to his car, which he'd parked on the road. As he roared off toward the docks, Hal turned back toward the kitchen to prepare for the day, wondering what LaRoux might do next.

Knapsack well packed and on his back, he walked briskly to the wharves. Despite his preoccupation with his various dilemmas, he noted what a glorious morning it was, sun clear, no fog, slight breezes turning the trees to many shades of green. And he loved the color of the ocean, light blue toward shore, deep blue at the horizon. Mulling over his talk with Maria and her presence in the cabin of the "Mary Q", he felt light as a feather despite his heaviest L. L. Bean boots and the menacing shadow of Oliver LaRoux.

As he approached the wharf carrying Maria's breakfast and their lunches, he saw that LaRoux had discretely parked on the public road and sat in his red convertible with the top down. But Hal pretended to pay no attention, walking directly to his truck to get a buoy he wanted to add to a trap. As he headed for the baitbarrels, he decided not to roll them out but scrimp by with the remains of bait already aboard. This would save stopping at the wharf and protect Maria. He vanished into the baithouse, pretended to fiddle with gear on the wall, then ambled toward the dock ladder ever so casually where he intended to climb down to his skiff. When he'd descend the ladder a couple of rungs, LaRoux appeared directly over his head, a lobster stave in his hand.

"Look you, young punk, I'll bash your head in if you don't bring that boat in here so I can see if Maria's aboard."

Hal continued to descend the ladder, out of reach of any stick LaRoux might swing, but he retorted, "Mr. LaRoux, you're trespassing and you're in dangerous territory. If you don't go away, I'll call the deputy sheriff on my CB and ask him to arrest you. If you're smart, you'll leave." All the while Hal was retreating via steady but quiet rowing.

As he approached the "Mary Q," he saw LaRoux get into his car and zoom off toward Statesville. He was glad that Maria stayed under cover, but didn't know how much she'd seen.

Climbing aboard with their meals, he knocked at the cabin door and she let him in. They embraced before he set the breakfast before them. She asked, "Hal, are you all right? I saw La Roux on the dock with that stick. What did he say?"

Hal recounted the story of his confrontation, the words they'd exchanged and his warnings about trespassing.

"Did you tell him we we're in love and are going to live together?"

"No, dear, I parried his questions and volunteered nothing. Frankly, I was fearful he might rush me and I'd have to throw him into the water as you did Walt Carmen!"

"Imagine, entertaining *him* with *my* karate lessons."

"Maria, I don't think this is over. From what you've said, he's persistent if nothing else. If he comes back and talks with guys like Digger and Sal and they downgrade us, he'll think we're cowards and harass us the more. But today we've gotta be careful. You stay here in the cabin until we get to the far side of Jocks Rocks and he can't see us from any shore angle. We'll fish until noon, then decide how we handle afternoon and evening. I'm inclined to present ourselves to the gang as we always do, competent fishers. What d'ya say, dear?"

"Aye! Aye! Captain Hal, what're your orders!"

"One more kiss before I start up the engine!"

Careful to avoid her injury, Hal kissed her warmly then asked, "How's your back?" He noticed she'd changed into her heavy fishing shirt and seemed to be wearing no bra.

"It could stand some more ointment," and she lifted the back of her shirt as he applied the soothing salve.

"Egads, Maria! It looks pretty raw. Maybe we should go to the clinic?"

"Let's carry on, dear. I'll try to pay no attention today. We can check it back at the house."

With that she finished her coffee and muffins, and Hal climbed to the deck to start up the "Mary Q."

Their day went well. Bait remained in some of the traps so they tended only half of them. They had only one scare. When a Baysville boat headed toward them, Hal shouted across the cockpit, "Jack Small. Prob'bly coming to follow up his early morning phone call. He's no gossip. Let's be honest with him."

Sure enough, as he pulled broadside, he opened with, "Mornin' Hal. Hi, Maria."

In unison they replied, "Mornin' Jack, what's new?"

"That's what I came to ask you. Were any of the boys messing with your boat when I phoned?"

"Nope!"

"What happened? Your father was here looking for you this morning, Maria? Or at least he said he was your father."

"It was him, Jack. Can't tell you everything now, but he's a dangerous man. Had a narrow escape from him last night."

"Oh?"

"Thank you for calling Hal. You two saved my life."

"Glad fer that, Maria. I don't want to see you hurt. Told yer father nothin', figured it'd be best to keep a tight lip. But he talked with Digger and Sal. Anything more I can do?"

Hal spoke up, "Thanks, Jack We're O.K. for the moment. We'll call on you if we need more help."

Jack said, "O.K. See ya later," and he wheeled his boat toward the harbor and roared away, at full throttle.

When he was out of ear shot, Maria practically crawled to Hal's side, holding her hurt shoulder. "Did I say too much?"

"No, dear, but I think we've had enough fishin' for one day; let's go in to save wear and tear on your back and shoulder. Want more ointment 'til we can treat it properly?"

"Might give me some relief."

"Want to lie down in the cabin?"

"No, I'm fine standing here. I'll change at the mooring."

All went normally at the dock. Jack was the only one in, busy with his boat and gear and asked no more questions. Maria and Hal loaded their personal things, now in large green plastic bags, into the truck and drove home.

Hal took her immediately upstairs to the guest room, suggested she get comfortable while he laid out towels in the bathroom. As he was about to leave for downstairs, she said, "Hal, this is to be my room?"

"If you wish?"

"And where is yours?"

"Right here across the hall?"

"And your parents'?"

"Behind that door. Unused since dad died."

"May I see your room?"

"Of course, Maria, but I've not cleaned for a week." He put his arm around her waist and steered her through the door. She saw it as a typical man's room with a beautiful prospect to the sea, an antique gun hanging from a rack, several of Hal's drawings tacked to the wall with common pins and one of his poems between the pictures. She read it aloud:

>   Ease gently
>   into whale and lobster roads
>   red bell buoy
>   clanging warnings
>   thru fog and nostalgia...

> grey weathered farmhouses
> closing warm grandfather clock sounds...
>
> cozying family treasures and mem'ries
> rosebuds fresh'ning scents
> crossing land and seascapes
> 'tween dawn and twilight...

She turned back to him to exclaim, "How beautiful, Hal!"

"Your voice was music to my ears," he said, continuing to admire her finely sculpted face. When she also exclaimed how lovely his room was, he put his hand lightly on her neck, softly remarking, "I'm glad you like it, Maria. But before we get into the afternoon, I'd like to do something I've wanted to do ever since I first saw you six weeks ago."

Maria said nothing, but put her left hand into his, inquiring with her eyes.

"May I take your head between my hands and feel your face as though I were totally blind."

Maria took a half step toward him. Though he knew his hands were calloused, he looked into her soft eyes and gently plied her face as though shaping a vase on a potter's wheel. She looked at him for a long time, then closed her eyes, at which point Hal took her in his arms and kissed her again, feeling every curve in her body.

Finally he said, "Let's forget cleaning house and making traps today, Maria, and do the Sherlock Holmes thing you need to do. Want to rest a little, change, what?"

First, let's put some more ointment on my back, then I'll change and freshen up. I want to see what the damage is? Much to Hal's surprise, she took off her shirt, turned her back toward him and stepped to a full-length mirror in the hallway, leaving him almost breathless as she exposed her breasts. He'd observed that they were medium-sized, firm and beautiful. In fact, he'd never seen a woman naked since he was about five when he happened upon his mother as she prepared for bed.

Maria saw him watching her, smiled and casually brought her arms up to cover herself, asking, "Where's the ointment, Hal?"

"Here in my shirt pocket."

"Please put some on, Hal. It'll sooth the fire in my back. I'll still wear no bra unless you think that would shock Jerry when we go over to Barnsville to ask him about father."

"Hard to shock Jerry. He took a shine to you anyway."

Hal gently rubbed the ointment into her bruises and welts even as Maria dropped her arms to her side and stood facing the mirror where he could still see her full-fronted. After a bit, she said, "Enough," walked to her clothing bag to get a fresh blouse. He felt the urge to take her in his arms again, but shyness again overwhelmed him and he hastened to the stairwell, saying, "Come down, dear, when you're ready to go. You want to talk with Jerry before phoning your mother."

Twenty minutes later, they were off to Barnsville. She'd freshened up, combed her hair, put on something fragrant.   Hal could hardly believe they were again traveling together along his favorite piece of road past the public parking spaces and surf.   Hal broke the silence, "Don't know why, Maria, but I keep putting off your driving along this road. Prob'bly not good for your back today....though you sure wrestled those lobster traps powerfully this morning."

"Why don't we leave it for another day, Hal, there'll be plenty of time to show me. I'm contented just being here beside you, and she moved to the middle of the seat and lay her head on his shoulder.

"You must be tired after all that's happened."

Eyes closed, she said, "Sorta," and a few seconds later he felt her head jerk as she dozed off. He kept his right arm and shoulder steady, concentrating on the road.

Arriving in Barnsville, Hal knew right where to find Jerry, at his favorite spot, whittling. Maria woke and he helped her out under the steering wheel, and they walked slowly toward Jerry. Both were a bit apprehensive, but they intended to pump him for what he knew.

He greeted them in his usual friendly way, grinning as he addressed them, "Well, if here isn't our young lobsterman and lobsterwoman."

"Hi, Jerry," they said, in unison, shook his gnarled hand.

Hal went directly to the issue. "Jerry, we need help and you told me to let you know if I needed anything. Now I do."

"Anytime, Hal, what can I do?" Jerry asked, glancing at Maria, standing nearby, arms akimbo.

"Do you remember the first time you met Maria and asked if she knew Oliver LaRoux?"

"Yep, sure do," Jerry responded, scrutinizing Maria again and noting that she was wearing no bra, her firm nipples protruding against her thin blouse.

Maria spoke up, "Jerry, how much can you tell us?"

Jerry wrinkled his forehead a bit, paused, and asked, "Why d'ya want ta know?"

Maria replied, "Jerry, my father would probably kill me if I tell you what he did to me last nite, but he beat up mother and me." She turned her back to Jerry, lifted her blouse enough for Jerry to see the welts.

"Gawd, Maria, that's awful. Shouldn't happen to a dawg. Sure, I'll tell you what I know. If he tries to hurt me, all the boys of our gang know him and will beat him up if he as much as touches me. He's been comin' here for years, met a whole buncha prostitutes, had one steady for most five years. She lived pretty high off the hog until she got AIDS and died here a few months ago. Don't see how he could have missed getting AIDS himself. Always seemed like a dapper man; always in a hurry; almost drove ovah one of the boys right over there," he said, pointing to the corner of the wharf. "A Catholic friend ovah in your town, Maria, says he's a valuable worker at the plant, a good family man, always at Church week ends. Saw him come outa the confessional booth one day. I nevah told him the diff'rence!"

During the next twenty minutes he recounted enough details about Oliver LaRoux to hang him.

As they prepared to leave, Maria put her hand on Jerry's bare arm, then kissed him on his cheek, saying, "Thank you, Jerry, you're a dear. What you just told us may save my life."

As Hal and Maria turned away, he tossed a word after them, "Heeaah, you're a damned good sternwoman, Maria. Yer a lucky man, Hal." Both turned back to wave, looked at one another, smiling and holding hands.

As they rode back the crooked road, Maria kept her counsel until Hal asked, "What do we do now, Maria?"

"Gotta be careful, Hal. Put a rat in a corner and he'll most likely fight. But we both know judo, don't we, Hal!" She looked across the truck cab and smiled.

"Yeah, but do we want to get that close to him?"

"Probably not, but how long before AIDS gets him?"

"Who knows?"

"Better that AIDS gets him than he get us?"

"When do you ask your mother about your birth?"

"Hal, you had a good hunch in asking me if I were adopted, and I intend to pump mother for details when we get home to phone her...even if I have to tell her all that Jerry told us...but never reveal our source."

Maria rode silently around the crooked corners which Hal was "taking a bit too fast," she thought. "Trying to kill us first, dear?" she asked, as the tires squealed.

"Sorry, guess I got too excited."

She said nothing, snuggled against his shoulder, and put her hand on his thigh.

Once home, Maria went directly to the phone, knowing she probably had an hour before her father would be home. She was surprised when he answered, quietly hung up, exclaiming, "Oh Shit!" They looked at one another then laughed!

"He'll be out for the evening, you can be sure of that, Hal, so we'll call after supper. Hey, Hal, reminds me; I'm hungry. What're we going to eat?"

"Can you stand wieners and sauerkraut?"

"Sure, let me cook. This may be a day of many firsts, first kisses, first confessions of love, my first meal for you, lots of first things. Remember, I love to cook."

They were in the full swing of preparing a meal when the phone rang. It was the lady at Whaleback Galleries about the Art Contest. "Been tryin' to reach you all day, Mr. Daniels. We wanted to tell you that you got Second Prize in the show. Congratulations! Also, we have an opportunity to sell two of your drawings, the big clipper ship and the pair of dories...."

Hal covered the receiver and yelled at Maria, "Hey, we got Second Prize and they want to buy some of my drawings. Shall we sell?"

Maria snatched a glass from the cupboard and held it up in a mock toast, rushed across the room to kiss his right ear. Hal turned back to the phone, saying, "Well, we didn't discuss anything like that. What do they wanta pay?"

"A hundred dollars! and we'll take 40%?" the manager responded. "Is that arrangement all right with you?"

"Why, of course, go ahead. I'll come over to see you some afternoon in a day or two, O.K.?"

"Fine!"

"Terrific!"

Hanging up, Hal grabbed Maria low on her buttocks and lifted her off the floor, whirling as he did it.

"Careful of my back, Hal, careful."

He lowered her to the floor, adding, "Oh, I'm so happy, Maria, I forgot myself."

"When you lose that boyish enthusiasm, Hal Daniels, I'll leave you. That's what I like, spontaneity. Congratulations, dear! All your work over the years, finally paying off. Now, Hal, congratulate me for my attitude and speech!"

"No wine in the house, Maria, but let's drink a toast with some Miller Lite, O.K."

While Hal and Maria were in a celebrative mood, drank toasts to one another and their future together, Hal felt a shadow cross her face and told her so.

"Yer right, Hal, I can't get this whole episode with my father out of my mind. As soon as we finish the dishes, I'll try to reach mom again." Which she did, and got her.

After a few "*bon jours* " and "*comment t'alle tu's*," Maria got directly to the questions she burned to ask. The conversation was in French, but Hal recognized enough words and felt the power of Maria's insistence to conclude that they were discussing the crucial issues. And he heard enough "*Mais Mama's*   to know that Maria's mother was resisting as she had in the past. But after a half hour of conversation, he heard Maria's anger turn to compassion, heard her refer to her father's AIDS several times and knew she was refusing to tell her mother where she was. He also heard her promise to phone again in a day or two, and close with an affectionate "*Bon* " and a simulated kiss.

Turning to Hal, Maria explained, "Your hunch was right. I <u>was</u> adopted! Oliver LaRoux <u>isn't</u> my biological father. Seems that his brother got a girl pregnant in Sherbrooke, Quebec, at some kind of family reunion. At the time mom and he couldn't seem to have children so they agreed to take the baby when it was born. I'm the <u>it</u>...." At that point, Maria turned to Hal, standing beside her as she sat by the telephone, looked up startled, tears welling up in her eyes, "Did you hear what I just said, Hal, I'm the it...the <u>it</u>...the <u>it</u>... And all these days I've been struggling for my identity. And I'm only an <u>it</u>?"

Hal handed her his handkerchief and kneeled on the floor beside her, put his arm around her waist, and whispered, "Not to me, you aren't."

Maria leaned over to bury her face in his neck, and he held her close until he heard the grandfather clock strike eight and looked across the dunes where the grass was turning yellow in the late evening sun. The clock's ringing seemed to bring both of them to their senses, and Hal asked her to tell him more.

"Well, mom warned me that La Roux...and Hal, I'll never call him 'father' again...had been on a rampage all day, stayed home and drank a lot of whiskey, returned from Baysville to curse you and the lobstermen he met. She said he talked and talked about the bitterness Francos have felt for Yankees She said he made some terribly outrageous remarks and warned me to keep an eye out for him 'cause she didn't know what he might do next?"

"Did she think he might show up here to rough us up?"

"She didn't know. But she confirmed what Jerry told us. She's known for a long time that he had a mistress in Barnsville, prostitutes in Statesville and Portland. Friends told her, but she didn't believe it until she hired a detective to follow him. She finally stopped denying it, at least to herself if not her family and the world. She's seen the report from the hospital about his AIDS test and thinks his company knows. He left the report on his night stand, maybe by design, she doesn't know. It seems that they never did talk about sex or their relationship. I was surprised she told me, but said LaRoux had always expected her to perform automatically, no matter how she felt.   She said she didn't want to believe the AIDS report at first and now feels guilty she didn't warn me in case he was successful in raping me.   And now she admits he battered her many times when Kathie and I were out of the house."

"And is she worried about getting AIDS herself?"

"They've not had sex since Kathie was born!"

Hal listened, thoughtfully, then asked, "Tell me, Maria, why didn't she report LaRoux to the police if she had such goods on him? She knew he'd tried to rape you a couple of times."

"Old story, Hal, as I've said before...it's the big lie." Maria, too, looked off toward the dunes, turned back in a mood of reflection, almost whispering, "the big lie..."

Hal continued, "Did your mom say whether or not La Roux owns a gun?"

"He does, a .22 caliber pistol, but she knows he's a lousy shot. Went to a shooting gallery with him not long ago, said he couldn't hit the side of a barn."

"Maria, dear, what should we do now?"

"Be careful!"

"Meaning what?"

"Well, for one thing, we're sitting here in this lighted dining room, probably a clear target if he decided to ride by and try to plug us from the road. Maybe we should start by turning off the lights."

As she reached for the switch on the table lamp, both heard an automobile pull into the yard. Maria snapped off the light before the car engine stopped while Hal made a dive for the wall switch. They heard the car door slam, were aware that somebody with a flashlight was coming toward the house. Hal whispered, "Maria, hide in the bathroom there under the stairs." Then all was still as Hal ducked behind the door to the living room, caught a glimpse of Oliver LaRoux's face and watched the light go around the house, throwing huge shadows whenever the man passed a window.

Hal positioned himself to see the front door while remaining close to the phone. Next came a pounding at the door. LaRoux shouted, "Open up and come out, Daniels, come out and act like a man. Maria, I know you're in there, too. Do as I say or I'll kill you both."

At that point, Hal quietly lifted the phone, shading its lighted dial with his hand, and called Elmer Kruse, almost whispering, "Elmer, a man with a gun is trying to get into my house and is threatening me. He's dangerous. Come quickly and quietly, but come!"

Elmer resisted, whined, wanted more information. Somewhat impatiently, Hal's whisper turned into a low shout. "I'm not <u>sure</u> if he's the same guy on the wharf this morning. Look, Elmer, this is an emergency! We can get into those details later. Come fast!"

Hal hung up the phone and Maria whispered from the john, "Is Kruse coming?"

"I hope so, but lie flat on the floor and stay behind that door, dear....please!"

Meanwhile, LaRoux continued to shout obscenities and threaten, his speech so slurred Hal knew he'd been drinking. When he smashed the large window in the front door with his pistol, Hal blew up and screamed, "You're making me mad, LaRoux. If you're smart, you'll get th' hell out of here!"

"I'm not leaving without Maria. If I have to kill you, I will." At that, he shone the flashlight through the broken window, flashed the beam side to side, then shot. Luckily Hal was well hidden; the bullet crashed into the dining room table, ricochetted into a picture on the wall, glass tinkling to the floor.

Marie asked, urgently, "Are you all right, Hal?"

"Yes, dear, but stay low and keep the door closed."

Hal kept hoping Kruse would arrive. He knew that LaRoux would see him if he moved. A thousand other thoughts flashed through Hal's mind: "Damn it, why didn't I keep my gun here...the one in my bedroom won't even shoot though it might threaten...and LaRoux can see the stairs so I can't make a break for it. Where in hell is Elmer Kruse, only three minutes away? Should I throw something at LaRoux, possibly a beer, and tell him it's a bomb? Should Maria tell him to go away...in French? No, Maria should not speak to him, that might make him angrier? Where's Elmer? Should I phone again? Is Elmer sore because I defied the gang? Could he have been paid off after all?"

Again, La Roux yelled at Hal, partly in French, partly in English. Hal didn't know all the words, but he was sure they were obscenities. LaRoux continued to hammer at the front door and poured three more shots into the dining room floor, each ricochetting into glass covered pictures or hanging plates. Again, Maria whispered, "Be careful, dear!"

"Please, Maria, stay back. You could get hit!"

"But, Hal, where are you? And where's Kruse?"

"I dunno, Maria. If LaRoux fires two more shots, I'm gonna rush him on the chance he's got no more bullets."

"Be careful, Hal," Maria urged. "Don't get hurt!"

LaRoux's fifth shot was higher, chipping the ceiling. It gave Hal got an idea which he relayed to Maria. Hearing the sixth shot, Hal let out a blood-curdling scream as though hit.

At that LaRoux put his shoulder to the door and crashed through it. As he entered, Hal gave him a karate chop in the throat and leveled him 'til he was crawling on hands and knees. In the half light Hal saw him lift his pistol again, kicked it out of his hand, yelled to Maria, "Come out and put the lights on, Maria, quickly."

She left her hiding place, flipped on the wall switch as Hal, put a full nelson on LaRoux. The light nearly blinded him, too. Though somewhat subdued physically, he continued to swear at Maria until Hal stuffed a handkerchief in his mouth and reached for a coil of rope in the hallway where he usually kept gear. Using no mercy whatever, he tied the man with bowlines and half hitches LaRoux had never seen before. Then he lifted the intruder bodily into a chair, tying him even tighter so he couldn't move. LaRoux tried to spit at them both, but couldn't get beyond the gag.

Maria looked on with admiration for Hal and contempt for the man she had called "Father" all her life.

As they heard Kruse pull slowly into the yard, Maria began speaking very calmly but deliberately to LaRoux. "Now I know the whole truth about you, Oliver LaRoux, about your rendezvous with prostitutes and mistresses all over the country, about your AIDS, about your adopting me as a baby to save your brother's reputation. And I'll never forget or forgive your attempts to rape me and what you did to mom at the house last night, how you threw her into a corner like a rag doll, and the beating you gave me. Just look what you did to my back," and she turned to lift her shirt enough so LaRoux could see.   As she spoke, she raised her voice into a scream. "Nobody should treat a dog as you've treated me since my sister was born. I couldn't figure out why you'd want to rape or disfigure your own flesh and blood. But come to find out, I am <u>not</u> your flesh and blood! And I'm <u>not</u> a part of your blessed Church with its ritual and hypocrisy. A great system and no better than the Protestant ones you detest. Go screw around then purge yourself at confession. Great for you and your soul, but how cruel for the memories and flesh of your family!" She paused as Hal looked on with a sense of justice being done and heard Elmer Kruse walk slowly up the walk.

Hal asked, "Shall I let Elmer in or do you want to say more to Oliver LaRoux?"

"Just one thing more, Mr. LaRoux!" Marie exclaimed. "You can take your worthless superiority and empty Franco morality and you can burn in hell for all I care. I hope you get put away for a long time, AIDS or no AIDS." With that she took a step toward him and slapped his face with all her might, stepped back and screamed, "I should have given you a karate chop instead, so take my slap as a gentle message! And I'm glad I'm out of your house and here with Hal Daniels. I love him and shall marry him. He's the kind of man who makes a person glad to be alive and human. You've <u>never</u> been that kind of man." She then signaled Hal to let Elmer Kruse in.

Hal's first words, "What'n hell kept you?"

"Sorry, Hal, but my wife had our community skunk in the house, wanted me to get him out."

"Well, Elmer, we gotta skunk here, too, and we got him tied up like Gulliver. He shot a round of bullets from this pistol into my house as you can see from the damage to my table and walls. He smashed my front door window and broke in to attack us, threatened to kill us both. Maria and I want to charge him with anything and everything that will put him away for a long, long time. We know he has AIDS so take whatever precautions you want to take. But give him no mercy. Maria and I will testify against him."

Not only did Elmer Kruse look up in surprise to find Maria, but by this time half the lobstermen of Baysville were crowding through the door to see what was happening.

Walt Carmen puffed, "Heard th' shots.... what's goin' on heeah?"

"Come in, boys," Hal invited. Along with Walt there was Digger, Ray, Sal, Jack and others from the gang.

Jack quickly spotted LaRoux tied to a chair and, turning to Maria, remarked, "That's the guy on the dock who said he was your father."

"That's right, Jack. He WAS my father until today when I learned that he and mom adopted me when I was a baby. The last twenty-four hours, he's been preoccupied in beating me up."

"Is that true, Hal?"

"Yup. Show the boys your lower back, Maria." She turned and lifted her blouse. Cries of "UN-real! UN-be-liev-able! Ouch!" went up from the group.

"So what're you going to do now, Maria?"

"Well, we're filing charges against LaRoux and hope to put him away until his case of AIDS does him in. Elmer, no bond for him; he's too dangerous to be roaming the streets."

Hearing such scary comments, the lobstermen moved back almost as a man.

"Oh, he can't hurt you now," Maria continued. "Meanwhile, you know Hal and I lobster for a living...although you guys haven't helped us much. Hal, do you want to give them the news?"

"Shuah, Maria and I will marry soon, right Maria?"

"Absolutely!" She gave Hal a hard hug and warm kiss. At that the men, including Elmer, applauded wildly. LaRoux just stared.

Jack Small cut across to Hal and Maria, shook their hands, gave Maria a peck on the cheek, remarked, "Gettin' kinda close in heaah, guess we'd bettah go home, fellahs. Unless, Elmer, ya need help gettin' that scoundrel to the county clink?"

"I could use one or two of you for a couple of hours. Will see that yer deputized and paid."

Jack said, "I volunteer." Sal, a bit sheepish and expressing apologies to his high school classmates, also volunteered.

Kruse and Small asked Hal to untie LaRoux from the chair while they held a gun at his head. Hal obliged, and they dragged the prisoner to the patrol car screaming against the gag and squirming against ropes that were cutting his wrists.

When they'd left, Hal looked at his watch just as the town fire whistle blew the nine o'clock curfew. "You know, Maria, I'd almost forgotten. I think there's an old bottle of sherry down cellar. I'll go get it so we can relax."

He found the bottle his grandfather had hidden, poured two glasses. They sipped the sherry slowly, then she steered Hal to the living room, pushed him down onto the sofa. She sat across his lap, looked into his face, kissed him, nuzzled under his chin. He was silent for a long time, savoring the smell of her hair. He stroked her forehead, neck and bare arms, felt her go limp on his chest, finally whispering, "Let me put you to bed."

She didn't answer so he carried her upstairs to her bedroom.

Chapter 10

Many Firsts, Many Climaxes

As Hal laid Maria on the guest room bed, she suddenly sat bolt upright, screaming, "No, don't do it! Don't do it! Don't!"

Hal grasped for her hand, pushed her gently back against the pillows, stroked her cheek, took her in his arms and whispered, "It's all right, Maria. It's all right. You're safe. You must have been dreaming."

"Oh, Hal, that was awful. Father...that is, Oliver LaRoux was chasing my sister with a pistol after ripping off her clothes."

Sitting on the edge of the bed, Hal spoke reassuringly, "Maria, you're all right, dear, you're all right."

Eyes somewhat glazed, Maria first pushed away, then reached up and grasped Hal around the neck, pulled herself into his neck, whispering, "Oh, Hal, I'm so glad you're here."

Both remained silent for many moments. When he lowered her back onto the pillow, she began to speak softly in short breaths. "Hal, we've worked together for only a month... but it seems like forever....you could have taken advantage of me....fault me for my mistakes...but you haven't. I've seen you peeking down the front of my shirt...and at times felt as though your eyes were reaching into my soul. I like your bashful ways...even last night on the boat...it's one of your strengths, like the gentle lines on your drawings. And when you've touched my hand or arm...or helped me with a trap or rope and bumped against me on the boat, I could feel my skin tingle. I wanted so much to hug you...be hugged...but you used no force. My trust, dear Hal, is stronger because of what the macho boys call your weakness. I do love you and want to hold you tightly...I don't think you'd ever hold me against my will."

Maria paused as Hal held her face in his hands, "Shhhhh, dear, maybe you need sleep to soften the pain you've suffered?"

Maria put her finger across his lips, "Please let me continue, Hal. I want you to hear what I'm feeling and thinking. I want you to love me for what I am and not just because I'm a woman...and I think you do. I love you because you're never pretentious even in your strengths. You're hard working but sensitive...you have a gentle toughness that makes me want to crawl into your skin and be you. Your silence speaks volumes. You've drawn, you've written beautiful poems like the one on your wall, you've overcome the machos because you care." Again, she paused and he stroked her bare arm, kissed her again.

She pushed him away but gently, spoke softly, "Hal, I've never had any boy friends. Maybe I hoped fate would bring you to me? And I feel the need to tell you everything. Despite Oliver LaRoux I've been touched only once, when a cousin and I were fooling around in a haystack at a family reunion in Quebec. I was eleven. Three or four of us were rolling in the hay. One cousin, a year or so older than I, made me grab his stiff penis, then he tried to stick it between my legs. I knew almost nothing about sex, but I knew that what he was doing wasn't right. I'd just started menstruating and was wearing a Tampax. Mother had had presence of mind and courage enough to prepare me that much. When I resisted his penis, he stuck his finger into my vagina but took it out fast when he discovered I was full of blood. He knew nothing about female puberty and was scared. I was scared, too. He then held me down while he masturbated all over himself. We never told anybody, but it frightened me. I wanted nothing to do with boys or men, spent my energy in outdoor sports and learning the martial arts. Only after high school did I learn much about sex during nursing training, reading novels and watching HBO. But even that was pretty remote and theoretical. So, Hal, here I am wanting sex with you but still very innocent, maybe as much as you?" And she smiled through tears as she looked up at him, running her fingers over his face.

"And, now?" he asked.

"Now," Maria said, "I'm <u>not</u> sleeping alone in this bed when we can be together in your double bed." She reached up and pulled his face and body down against hers and they lay there a long time. He broke the spell, rose to sit on the edge of the bed, quietly observed, "Maybe we can learn together 'cause I don't know much either, only read and watched things on TV. But for a long time I've fantasized beginning in a beautiful forest, lying in ferns and flowers, under a full moon. Have you ever imagined anything like that?"

"Not really, Hal. All the forests that I know are full of mosquitoes or snow. And tide charts told me yesterday that the full moon isn't due for another two or three weeks. But I have an idea how we can manufacture a full moon...but only after we've had a shower together. Come, let's take one."

Together they went to the bathroom, and Maria began to undress them both. Hal, now fully aroused, looked down and started to apologize, "I'm sorry, but I can't help...."

Again, Maria 'shhhhh'd' him, putting her fingers across his lips, whispering, "Of course you can't help it. That's what it's for and about..." She continued to strip him to his shorts. Meanwhile, she took off her blouse and jeans, getting down to her panties.

She then started the shower, feeling the temperature with the back of her hand so they'd not be burned. Stepping in herself, she reached out and quietly guided Hal in. She handed him a bar of soap and began to work up a lather, telling him to turn around so she could soap his back. When he turned to face her, she put some suds into his hand and asked him to wash her arms and shoulder blades. In a moment she turned him around again, stripped off her panties and slid his shorts off. And so they progressed, step by step, until they were both fully soaped and massaging one another gently. They only took precaution with her back. She then got down on her knees and washed his legs, ankles and feet, then asked him to do the same for her. She took his hands and ran them along her upper thighs until they reached her pubic zone. And finally, she took his penis in both hands and flattened it against her belly even as she pressed her breasts against his chest. In that position they stood under the warm and soothing water for a long time...in fact, until Hal's erection was gone. She laughed a little, "See, I told you everything would be all right."

At that she turned off the water, handed him a towel and asked him to wipe her down. That finished, she dried him. All the while they laughed joyfully. But whenever Hal began to speak, she laid her fingers across his lips. Once dried off, Maria led Hal to his own room, pushed him gently into bed, then jumped in behind him. But just as quickly she bounced out, noted it was totally dark outside, closed the curtains, and went to Hal's crooked-necked, bullet-like lamp at his drawing board. After she snapped it on, Hal lay on the pillows, admiring her strong upper body, beautiful breasts and buttocks. This, of course, aroused him again. She turned the light toward the ceiling, marked by a tight bright circle, laughing, "There, Hal, there's your full moon," and jumped back into bed, this time straddling him while smothering his mouth with kisses and moving her breasts along his chest and face so he could kiss her nipples. Within seconds they were fully coupled and moving together in rhythm as she whispered, "Hal, we must do it this way tonight because I can't lie on my back. But I do promise, some other time..."

Hal said nothing, but kept his eyes on his full moon and her ample and undulating buttocks and hips, moving to a crescendo which made him totally unconscious of the rest of the world. He heard her swoon several times before both his moons exploded and they lay quietly in one another's arms, both conscious and unconscious. Only the striking of the grandfather clock at eleven awoke them, Hal speaking first, "Thank you, dear, that was worth an ocean of pain."

As she rolled over on his chest again, "Thank you, too, Hal. For a few seconds, I thought I'd died and gone to heaven."

"In a day of so many firsts, Maria, that need not be the last; we'll have lots of heavens to go along with our hells. With that he again smothered her face, hair and breasts with kisses before stopping short to chuckle, "But do you know what I just remembered I forgot."

"What?"

"I forgot to set the alarm."

Maria laughed and poked his ribs. "No need, silly, tomorrow's Sunday!

By force of habit both awoke at dawn. Hal watched silently as Maria slipped out of bed, crept softly to the window, pulled back the curtains, and stretched against the full flush of golden light pouring from the east; again he was all eyes as her full nude form cut a sharp silhouette against the dawn. "I must sketch her," he thought. He then followed her as she tip-toed quietly around the foot of the bed and went to the john. Every creaking board in the old house seemed magnified, also the squeaking bathroom door, the sound of her peeing, and the flush. It was such music to his ears, something he'd missed since his parents died.

When she returned to the bedroom, she saw he was awake, stripped off the lone sheet that covered him, then jumped on him again, tousling his hair and pretended to resist his wandering hands. But finally he pulled her down to him so they were nose to nose.

"What're we going to do?" she whispered.

"Do it again if you wish."

"I wish, but then what?"

"Sleep til noon."

"Beautiful idea."

When Maria next awoke, Hal was gone. She could hear him thrashing around in the kitchen, smell coffee and bacon and hear the teakettle whistling. She remained quiet, head on one elbow, waited to see what might happen. Within a few minutes he appeared in his undershorts at the bedroom door with a tray and two breakfasts laid out side by side.

"I can't believe it!" she exclaimed.

"Better enjoy, enjoy," Hal chuckled, it might be another first, maybe a last. You'll never get this kind of service mornin's when we're heading for Jocks Rocks before dawn!"

Maria wrapped the sheet tightly across her breasts, lifted her arms to grasp Hal's neck as he leaned over to lay the tray on her stomach, said "How can I tell you how much I love you?"

While eating side by side, Hal asked, "Well, dear, what're we goin' to do today...the day after the night before?"

Steadying her coffee, she pecked him on the ear.

"Is that an answer?" he asked.

"It will always be an answer."

"Then, what?"

"Hal, one thing I've gotta do."

"Bet I can guess."

"Are you a psychic?"

"Sometimes!"

"What?"

"Go see how your mom's taking all this."

"You guessed it."

"Makes sense, Maria, after all you were close to her."

"Yes, Hal, but I'm still kinda mad she didn't level with me about lots of things, especially the adoption. But I suppose I have to forgive her. No question, Oliver LaRoux intimidated her all their married lives."

"Also, I suspect you've more clothes to bring home."

"You mean you don't like me like this?" she teased, throwing back the sheet.

"Of course," Hal said, leaning over to kiss her breasts, belly and groin, though somewhat awkwardly.

"Ow, that hurt! Your beard's sharp as sandpaper."

"I'm sorry, Maria, I'll shave in a moment. But wanta go see your mom?"

"Of course, dear, if you'll go with me. With LaRoux gone, I think it's safe."

"Unless he's broken out of jail."

"Gawd forbid! Bet the jailers are still working on your knots!"

They were dressing one another when they heard a heavy knock. Hal tiptoed to the window above the porch door, peeked out to see who it was. "Egads, Maria, there's a Statesville GAZETTE car and what must be a couple of reporters. One has a flash camera. Do we want to talk with them?"

"Not really, Hal, unless you want to?"

"Then how do we get rid of 'em? They can see our truck in the yard so know we're here?"

"_Our_...truck...Hal?"

"Of course, what's mine is yours!"

She was pleased with Hal's remark, giggled a little, agreed they probably had to talk but should volunteer nothing, only answer questions.

Hal yelled out the window, "Be right down."

Once the reporters flashed their credentials and were inside the door, they behaved much like the proverbial camel in the tent, taking pictures at random. Hal finally steered them to the sofa. The one with the pad spoke first. "We've talked with Elmer Kruse, been to the prison and taken pictures of Oliver LaRoux, but he won't talk. Some of your buddies hanging around the wharf told us part of the story. They say you were victim of an attack, Maria, and that you are a hero, Mr. Daniels. What do you have to say?"

"I don't feel like a hero. First Maria was in trouble, then we both were in trouble. We did what we had to do."

"Like dodging bullets?"

"Not really, but he cut up the dining room somethin' awful, as you can see for yourself," pointing to the dining room.

"May we take more pictures."

"No, please don't. Keep details to a minimum. We don't want to do anything that could hurt Maria's mother and sister any more than it has already.

The reporters left as quickly as they'd come, leaving Maria and Hal to wonder what the paper might say. In a moment or two one reporter rushed back to shout through the screen door, "Hey, Mr. Daniels! Are you the same Harold Daniels who won a prize in the art show yesterday?"

Hal said, "Yes, but how'd you know?"

"Saw it in our mornin' paper; haven't you seen it?"

"No but we will."

As the company car backed out of the yard, Hal dashed across the road to fetch the rolled up Sunday paper from the white box. Within moments he and Maria were reading the account of the show. The article was very generous in praising Hal's sketches, suggesting that he might have taken first prize if a veteran painter hadn't entered the contest.

Maria grabbed him around the neck and hugged him tightly, "I'm so proud of you, Hal, so proud."

"But you deserve honorable mention, dear, you helped select them. I wonder if any of our old teachers will read this?"

"If they do, I hope they feel guilty."

"But, Maria, maybe they helped me by opposing me? If they'd not tried so hard to stop me from drawing, maybe I wouldn't have persisted?"

"You do come from stubborn Yankee stock, Hal Daniels, but I'm so glad I'm joining you to make new stock."

"You're planning children already?"

"Of course, dear, don't you want some, too, especially boys to carry on the family tradition?"

Hal picked her up in his arms and started up the stairs, but she squirmed away, running toward the dining room table, defying him to catch her.

When he saw her start between the wall and table, he grabbed her arm. "Careful, Maria! Don't cut your feet on that glass! I'll fix you tonight!"

"I hope that's a promise," she laughed, hardly containing herself with happiness. She hugged him again, and they left the house hand in hand. "Better go to Statesville in the truck, Maria, in case you want to bring back all your stuff."

"S'pose the Omni's still there?"

Hal looked toward his spare garage, responding, "Lock's still in place."

The day was hot but gorgeous. As they rolled along the highway, Maria babbled away, rehearsing the speech to her mother. Hal listened but said little. He felt her preoccupation so paid attention to the Sunday traffic, filled with tourists. As they pulled up in front of Maria's old home, Hal asked, "Nervous?"

"A bit, but I'm counting on you to help me."

"And that won't upset your mother?"

"I don't really care, Hal, she deserves to have some guilt, and you deserve to be involved."

Maria had phoned to set a time for arrival. And sure enough her mother answered the doorbell promptly. But something unexpected happened. Kathie was at her side.

Maria kissed her mother rather formally on each cheek, then flew into her sister's arms, exclaiming, "Kathie! I didn't know you were home."

Her sister was both cool and rigid, said almost sarcastically, "Thought I'd better get home to check the pieces after hearing the ll o'clock news last night."

Maria turned to Hal, took him by the hand and excitedly said, "And I want to introduce you to my fiancee, Hal Daniels."

Both Kathie and her mother gasped, shook Hal's hand very correctly, neither offering congratulations.

Maria observed this coolness, blurted, "We need to talk!"

During the next two hours, the four of them chattered about family affairs, her mother's state of mind and new plans to return to work, Kathie's future, reasons for Oliver LaRoux's behavior, the why of his trying to rape both daughters, the timing of all this in their lives. Hal was somewhat embarrassed to hear the three women "wash their family's dirty linen;" yet, he kept telling himself it was important to stay cool and get a full picture. Several times Maria grabbed his arm or hand, and he could feel the perspiration running from her hand to his. Both her sister and mother insisted upon seeing Maria's back, so she stood facing Hal and took off her blouse, causing them to gasp. He wasn't sure whether they were gasping at Maria's raw back or the morality of Maria's disrobing, especially without a bra. After she'd put her blouse back on, Maria put her hand on Hal's shoulder and said, "If he'd not put first aid ointment on it so soon after it happened, it would be worse. She leaned over and kissed Hal on the forehead as Hal noted the mother and sister looking at one another curiously.

When they got around to discussing marriage, both mother and sister bombarded Maria with questions, in French, wanting to know when, where, how, all the details. Her mother was most pointed, "Of course, you'll be married in our church?"

"No, Mom, I never want to go into that church again. You and Sis can go as much as you want, but to me it stands for 'evil'. Hal and I haven't discussed this yet, but I'm making the ocean my church. I find more that's sacred there than in my old church. Sorry, but that's the way I feel."

"And you won't bring up your children Catholic, then?"

"No, definitely not! Hal knows how I feel about all this."

The hour getting late, Maria looked at her mother and said, "Mom, we've talked about the life insurance policy you have on me?"

"Yes," her mother responded, unhesitatingly.

"And what's it now worth?"

"A little over $4000. I cashed it last week, opened a personal account at the bank. I sensed Oliver was about to do something more stupid than usual, but I didn't want to get into a shouting match with him about it."

"And the money is now mine?"

"Of course. I'll transfer it to your account tomorrow. I wouldn't go back on my word. You've been a good daughter up to now. I don't like your marrying somebody not our kind or leaving our church; but if you love Hal, you'll get my blessing. She turned to Hal to ask, "Incidentally, do you have any money put away, Hal?"

"Not much, Mrs. LaRoux. Most of it's tied up in my lobster boat, house and fishing gear."

"So you're not marrying Maria for her money?"

Maria laughed and Hal replied, "We've never discussed her finances, only what I was paying for work on the boat." Hal felt like telling her "where to go," but Maria put her hand on his neck and whispered, "It's O.K."

"One more thing, Mama," and Maria again broke into a lengthy French discourse with both her sister and mother. After this barrage, Maria turned to Hal to explain. "I have a small dowry consisting of blankets, sheets and coverlets, things mom has made or bought over the years. And they're all in a cedar chest in my old room. We'll take them home this afternoon. So let's go get the chest along with the rest of my clothes. Oh! one other thing, Kathie. Are you still planning to go to the University this fall?"

"Of course, Maria, you know how much I want to be a vet. And I've saved enough to get me through two years if I stay at home and commute. And with father, I mean Oliver LaRoux, out of the way, I guess it's safe enough, don't you think?"

"Yeah," Maria said, almost absent-mindedly, looked at Hal to ask, "Can we help her if fishing stays good?"

Hal nodded with a simple but firm, "Definitely!"

As they stood up and prepared to collect Maria's things, Kathie asked, "And may I kiss my future brother-in-law?"

Maria smiled as Kathie crossed the room to give both Maria and Hal warm hugs. Mrs. LaRoux maintained her formal posture, shaking hands with him very correctly.

Rolling along toward home, Maria's possessions firmly secured and covered with a tarpaulin, Hal stopped at the crossroads leading to Barnsville, got out of the truck and walked around while Maria watched him with some curiosity. He opened the passenger side of the pick-up and started to get in. "Darling, this has been a week end of many firsts. It's time you drove the crooked road from Barnsville to Baysville. Who knows what idea might jump out of the forest at us?"

Maria was a little slow sliding into the driver's seat, continuing to favor her sore back. Hal noted, reached around to get a corrugated seat pad to make her more comfortable. With little fuss, Maria started them off toward home via Barnsville, noting how easily it drove.

"Automatic transmission, power steering just like your Omni, dear."

As they approached Barnsville, Maria asked, "Want to stop and see Jerry?"

"I'd really like to get you home and to bed to rest after all of these strenuous events, but maybe we'd better see if he's at his whittling post and tell him the news first hand. We owe him that courtesy, Maria, though I imagine he's been hearing rumors as loud as seagull cries all day!"

Sure enough he was at his favorite spot whittling model seagulls as they pulled within four feet of the old wooden seine tub on which he sat. He looked up, waved with his knife hand, laid down his carved piece, and stood to greet them.

"CONGRATULATIONS to you both. Jack Small was over to borrow some bait this morning, gave me a play-by-play account. I figured you'd marry sooner or later, but what a story!"

Maria spoke softly, "Jerry, you've been such a friend, we'll name our first child after you!"

Hal looked amusedly at her, then to Jerry, "And the kid isn't even conceived!"

She glanced up amused as if to say, "How do you know?"

Driving home, Maria shared that thought with Hal, and his eyes opened wide as though to say, "Can it be?" They also hatched some of their wedding plans. "Hal," Maria began, "I wasn't kidding when I told mom that the ocean was now my church. Could we get married on the 'Mary Q?' Let's find a Justice of the Peace to marry us on the boat or even by CB? And let's do it on a moonlight night to fulfill your fantasy about the full moon and love? What do you think?"

"Fantastic idea. Maybe we could combine it with something I've been thinking about. Let's get married on a Thursday night when the moon is full later this month, then spend Friday, Saturday and Sunday at Jewell Island, just along the coast. I was there once as a boy, and it has a beautiful lagoon where we could anchor the boat, be safe and calm and fill our days and nights with love." He reached over and touched the bare knee of her driving leg.

"Careful, Hal, or you'll have me off the road. I'm 100% for our joint plan. Tomorrow afternoon, let's go to Statesville to work out details about licenses, the Justice of the Peace, etc. Now we've gotta get home to rest for tomorrow's fishing."

"Rest?"

"Rest!"

Chapter 11

Choice, Chance and Will

Five o'clock came early that Monday morning. Maria and Hal were both "sensitive" and "sexually sensible," as they frankly noted, "hitting the deck" as soon as awake to prepare for a routine fishing day. They'd agreed to share home chores and moved quickly to do them. Hal got breakfast and lunch; Maria checked NOAA radio for weather reports and tides, gathering a coil of rope and a clutch of styrofoam buoy materials to throw into the truck. Hal suggested that she check her car again for safety's sake. They knew they had to talk about their cars, go grocery shopping, and arrange formalities for their wedding, now set for the 20th, the night before the full moon.

As they drove onto the wharf they were surprised to see so many of the Baysville gang gathered in a knot in front of the baithouse. Before leaving the truck, Hal whispered to Maria, "Wonder what mischief they're up to now?"

Maria replied, "Don't know, dear, but please be careful," and gave his arm a squeeze.

As they walked side by side toward the group of men, somebody shouted, "Hip!" then the entire group yelled, "Hip, hip hooray! Hip, hip hooray! Hip, hip hooray!"

"What's this?" Hal asked of Jack Small, standing at the forefront of the group.

"Hal, we just had a rump meeting of the Association and all agree on something for the first time in years. You and Maria deserve our cheers. So we're cheering for you! We congratulate you both for what you did for this town last Saturday night. You brought us together, and we want you to know it. We don't need enemies within with people like Oliver LaRoux out there. And I speak for the Baysville gang. We're sorry for the hard time some of us have given you."

Hal felt tears coming to his eyes, saw that Maria was also beginning to cry. Reaching across the tiny space between them, he pulled her head to his shoulder. More cheers. "Speech! Speech!"

Together they mounted a nearby lobster trap and Maria spoke first. "As a newcomer I want to thank you. We'll tell you about our wedding plans as they develop."

More cheering.

When Hal began to talk, somebody from the back of the crowd yelled, "Is she a nice lover, Hal?"

Jack spoke sharply, "O.K., John, none of that! Let Hal speak."

Hal continued, but nodding to John said, "Use your own imagination," and that brought laughter from the group and a dig in the ribs from Maria.

Hal went on, "We're very much in love, and we intend to hold up the Daniels and Baysville traditions. Just so you know that we're going to work hard, we're going to Barnsville tomorrow where Jerry Moulton's arranged for us to buy fifty more traps. We know you like good competition, but we've got the legacy of this town to uphold.   Are you with it?"   This led to more cheering before the group dispersed to their boats and jobs.

Roaring out to Jocks Rocks, Hal reviewed Maria's plans to spend some of her insurance inheritance to buy fifty traps, Hal noting, "Are you sure you want to do that, Maria, and not buy clothes or something else you want?"

"Your love is clothing enough, Hal, I wasn't kidding when I told mom the ocean was my church."

"You know, we could lose fifty traps in one storm?"

"Yes, Hal, but we've agreed to have 500 by Labor Day. Next winter we'll make another 50, 100, whatever you think best to make a decent living. So long as we're working together, I don't care what happens. You've always been up front about risks. And please, dear, let's never argue about money."

Since they were almost to the first trap, Hal held her tightly around the shoulders, asked about her back and said, "Hate to let you get too far away!" She pecked him on the cheek, twisted away and balancing against the rocking boat shuffled along the deck to the stern to her fishing chores.

The day went routinely. Catch wasn't too good, but they caught enough to break even although rumor was rife that boat prices were going down that afternoon. During their mid-morning lunch break they talked excitedly about the wedding, made "To Do" grocery and general shopping lists. Staring to the starboard into the bright eastern sun, they noticed that Digger seemed to be having trouble with his engine. Hal called him on the CB to see if he needed help. He was a bit distant and cool, but in a chagrined tone said, "Thanks, Hal. Water in the carburetor I think. Guess I can make it O.K."

"I'll leave the channel open, Digger. Sing out if you need help."

Maria moved to the CB, snapped it off momentarily, remarking, "And he was once your enemy?"

"Yes, Maria. With time, let's hope for peace."

"Positively, dear, but let's not drop our guards too far."

"Right," Hal replied, matter-of-factly as Maria turned the CB back on.

Roaring back to the docks with the engine wide open, Hal asked Maria to take the wheel while he cleaned the decks. When he returned to her side, she said, "Been thinking, Hal. Have a question: I told you the ocean's my church. Would you mind so terribly if I cleaned and reorganized what for me has become sacred space, the cabin of the 'Mary Q'?"

Hal put his arm around her. "Dear, the boat is as much yours as mine. We'll soon have papers to prove it. Do whatever you wish. Sorta hard to keep it clean down there, but if we keep the cabin door closed we can keep out flying seaweed, seagull doo-doo and other junk. How'd that be?"

"That would be as nice as this," and she pulled his head down to kiss him. "Hal, during the past twenty-four hours, whenever I've felt a nightmare coming on, I've imagined you and me in this cabin, you and me in bed. Warm and beautiful pictures to keep old ghosts at bay!" With that she relinquished the wheel, and he threaded his way into the harbor, taking a shortcut only possible at high tide on a calm day.

They spent that afternoon in Statesville, attending to the marriage license, finding the name of a Justice of Peace in the Baysville-Barnsville area, stopping into the gallery where Hal collected his hundred-dollar prize and proceeds from the sale of the two pictures (minus 40%), also learned that the manager had a prospective customer for a lobster boat sketch. She was not only complimentary but also wanted to know when he could bring in more. Maria went to the bank, discovered that her mother had transferred her insurance money, $4133.48 to be exact, into Maria's modest money market account. While Hal was grocery shopping she phoned to thank her mother and talk briefly, learning that her mother had found Hal a "pleasant young man" and that she wasn't as angry as she'd been when first hearing about him. "And, Maria, I liked what they wrote about you in the morning paper!"

"I'd almost forgotten, mom. What'd they say?

Her mother read one paragraph before Maria cut her off."

"O.K., I'll buy a copy before we go home. Hal and I'll have fun reading it together."

When she met Hal, she discovered that he'd bought a paper, too. They huddled on the front seat of the truck and read the front page article, looked at one another as Hal said, "Better than stopping one of LaRoux's bullets and landing on the obituary page!"

Maria clutched his arm, tightly.

On the way home via Barnsville, they picked up 15 traps. Maria wrote a check to pay for them. They made arrangements to pick up the other 35 the next day. As she passed the check to Hal, she smiled and remarked, "That buys me in!"

Hal responded, "So now, it's flesh, blood and money!"

"You're not kidding. And looking sidelong at the young fellow loading the traps, she said, "But, Hal, I have another question to ask you in the truck."

Rolling down the crooked road, with Hal at the wheel and Maria driving him crazy by rubbing her hand along his bare thigh, she asked, "Hal, have you ever thought of taking classes at the Portland School of Art?"

"Can't say that I have."

"Well, you're too good not to be better, and with the money I just got from mom and the $753.22 I had saved, we could certainly afford it....IF you want to. Maybe I could work on my bachelors degree at USM if we could get classes the same afternoon or night."

"You think I need courses, dear?"

"No, not necessarily. You have so much natural talent. But you could probably improve so long as you were careful not to get some dominating and stubborn teacher trying to make you over into their image."

"Let's talk about it some more, Maria. Aren't you also telling me that you'd like to get a college degree? Wouldn't that make you the most educated lobsterman, I mean lobsterwoman, around?" Hal smiled.

Maria chuckled, "Maybe..." paused, added, "but the idea would be to grow together. If you didn't want me to, that would be O.K., too, but some of my nursing courses might count. I guess what made me think of it was having a little money we didn't absolutely need....also your prize in the show."

"Maria, I've been mulling over an idea that relates to yours. Could I draw you in the nude? From what I read, schools test artists by their ability to draw the human figure. And I'd never seen a real live nude until I saw you. No matter how long I live, I'll see you silhouetted against the dawn. No wonder artists are inspired by so much beauty."

"Was it beauty or sex that inspired you, Hal?"

They joined in laughter as Hal brought his hand down on top of hers, still high on his thigh.

During the next two weeks they filled routine days with dialogues, plans, joy. Nights, too, sometimes sleeping little and loving much. But they paced themselves passionately on both bed and sea. Maria acted upon her intention to turn the boat cabin into a kind of shrine to their love. But she pled with Hal not to peek, also hold off loving her there until their honeymoon.

The night of the Fishermen's Association meeting was the night before their wedding, and it was memorable, too. Early in the session, Jack Small replaced

Digger Danes as president and swore in a whole new slate of officers. Jack handled the issues smoothly, speeding up the meeting by requesting delay on discussing the strike issue, resolutions on keeping Canadian lobsters in Maine pounds to regulate prices, trap numbers and the on-going conflict with the yachtsmen. Little did Maria and Hal realize what he was trying to do.

Hence, when Maria's application for membership came up, Jack called for the motion, got a second and entertained discussion. Both Maria and Hal were astounded when so many of their fishing colleagues stood to testify about Maria's and his competence as fishermen and their contributions to Baysville. After the motion to admit her passed unanimously, Jack announced, "I have two special loving cups, one for each of you. Come up here while I read the citations:

> 'To Hal Daniels
>   ...for furthering the
>   Daniels Tradition
> 'To Maria in honor
>   of being the first lobsterwoman
>   in Baysville.'"

Members of the Association applauded heartily.

Then Jack said, "I have another surprise. Will the Auxiliary Women please come front and center?"

They filed in briskly from the kitchen, each carrying a gift in one hand and a dish of food in the other. Maria and Hal realized what was happening, broke into tears and began blowing their noses. They could hardly believe how they'd advanced from goats to honored and accepted citizens. But Jack was not through reading from his notes:

"Hal, we confess our guilt for being so slow to see the wisdom of your hiring Maria as your sternwoman. Our wives, too, admit how hostile and jealous they were to see you out there at Jocks Rocks, Maria, when they felt that tradition and household chores kept them home. We hope this party will be our tribute to your courage and your marriage tomorrow. We discussed having sep'rate stag and bridal parties for you. But we decided it would be inappropriate because you've truly united us. Come on, you guys, break out the champagne and beer and let's have a toast."

Within a minute everybody had glasses in hand and even Digger looked happy as he spoke for the group,

"To Maria and Hal, may ALL your offspring
be fishermen and fisherwomen!"

Many shouted, "Heah! Heah!" and more cheers rose to the ceiling.

When the noise died down, Hal spoke. "Maria and I have just been discussing how to respond; we, too, propose a toast:

"To the Men & Women & Future
Children of Baysville!"

Jack introduced a combo and shouted, "Let's dance!" Maria danced with all the men as Hal looked on, not knowing how to swing...though she took him in tow toward the end of the evening and he began to get the feel of it. After opening presents, eating, socializing, and dancing until long past normal bedtime, Jack finally announced, "One more dance before we close and a final announcement: THE WHARF WON'T OPEN UNTIL 7:30 IN THE MORNING."

Much laughter.

When they got home, Maria threw herself across the bed, saying, "I'm emotionally exhausted, Hal. Could you have ever guessed they would do that?"

"It would never have happened for me alone, dear. You've made the difference in <u>all</u> our lives." And he lay down to caress her and watch her fall asleep.

They slept until 6:30 their wedding morning, and, as agreed, tended to "business as usual." Once aboard the "Mary Q" Maria took one last moment to prepare the cabin, asking Hal not to inspect what she'd done. Good naturedly, he said, "We now have a split captaincy!"

Maria said, "Of course not, silly, only for tonight!"

Back at the house after fishing they made final preparations for their cruise to Jewell Island. Maria had made enough casseroles to last them three days, prepared fresh fruits and desserts and had managed to keep Hal from snitching too many goodies. During the previous two weeks, he'd sat at the kitchen table, ever ready to serve as "go-fer" as she cooked up a storm. Consistent with their mutual wishes, she frequently worked around the house in the nude with the drapes pulled, and he caught her in various poses, wielding a broom or the vacuum, beating an egg, washing dishes as he wiped, having a beer. At one point when she reached in the cupboard for a sauce pan, he whispered loudly, "Hold it! Hold it!" She laughed and retorted, "My back just healed, dear!"

"Somebody, somewhere caught ballet dancers in various poses, Maria. I want to catch you in both graceful and awkward poses, too."

"Some French painter, we'll look it up and get a book about it. But, Hal, it would be more comfortable on the bed."

"Of course, any time!"

"Hey, posing I mean, nothing else!"

They laughed uproariously, took a long look at the kitchen clock, and she said, "Almost sunset, time to meet the wedding party at the docks."

"All three of them!"

"And no more?"

"No more! the Justice of the Peace, your sister and Jerry Moulton."

Dressed in clean jeans and denim shirts, they put their gear into the truck and drove to the dock which by agreement was vacant. Sure enough, there was the party; it was almost dusk, and the group climbed down into a borrowed dory so Jerry could row them out to the mooring. As they approached the port side of the boat, Maria happened to glance at the stern to see that Hal had changed the name of the boat to

THE MARIA D

She looked at him in surprise, almost tipping the dory over as she stood up excitedly, pointed and exclaimed, "Look! Look! You changed the name!" At that moment Hal reached over the side of "The Maria D," pressed a button and a tape recording of "Here Comes the Bride" burst forth. She was in full flood of tears by the time that Hal helped her aboard, gave his hand to Kathie and the Justice who said, "I feel like a fish out of water doing this." After some shuffling about, participants took their places, Maria and Hal with their backs to the cabin; and as hoped they faced the fulling moon as it came dripping, like a huge orange, from the horizon. By this time both Maria and her sister were sniffling.   Maria stammered through the tears to say, "This is all.....so.....over-whelming....the beauty of it all..." The harbor was completely still. Only a far-away dog's bark echoed across the water. Twilight was fading from orange to lemon yellow. All of a sudden, at a time predetermined by Jack Small and Jerry Moulton, the Justice of the Peace raised both hands, and every houselight in Baysville came on.

Kathie read a haiku that Hal had written for the occasion:

> Two buds become blooms
> as seas churn into combers,
> straight-line horizons...

Then the Justice routinely performed the ceremony that Hal and Maria had agreed upon. They exchanged rings, the Justice stressing the fact of their being on the good ship, "The Maria D," concluding with the pronouncement that they were now "man and wife." As they looked across to the wharf, every citizen of the little fishing village now stood with a lighted candle. Maria was happy, too, to see her mother with the group. Luckily there was no breeze and only a little flickering of the candlelights as Hal and Maria watched Jerry Moulton row Maria's sister and the Justice back to the dock.

Before Hal started up "The Maria D," he and Maria joined to wave and shout in unison, "Thank you! Thank you! See you Sunday night!"

Maria and Hal embraced and kissed; the citizens cheered.

Motor started, she released the mooring, and he wheeled the boat toward the open sea, turned on the running lights, and gunned it toward Jocks Rocks, moonlight square in their faces.

"You know, Hal, a billion dollars wouldn't have bought us a more perfect wedding. And now we're free, free for three whole nights and days."

He hugged her tightly against his right side, steering with the left hand, confident of their love, of their joy and their destination. "And we agree, Maria, to lie off Newtown Beach tonight and push on to Jewell Island tomorrow?"

"Whatever you say, Hal, you know the Coast better than I. Just anchor in smooth water if you can. I'm ready for our cabin anytime."

Outside their usual fishing grounds they encountered a heavy roll so Hal changed his course slightly, decided to anchor in a small harbor on the lea side of Gull Isle where the boat wouldn't rock so much. The moonlight was scintillating, each nearby wave picking up the light to paint brilliant splinters and streamers along the surface. Close by the water was filled with phosphorescence and they saw an occasional fish swim along. In the distance a string of lights etched the shore, houses and old hotels looming as long black silhouettes.

When they finally dropped anchor and Hal shut off the motor, Maria let out a sigh. "What a relief! First, dear, how did you manage to change the name of our boat without my seeing it? After all, we were back at the mooring at noon."

"Used the prize money to buy gold and black letters, made a deal to have Jack Small do it this afternoon, really quite simple since mother's name was close to yours! Remember I once said you might be honored this way, too?"

"At that time, Hal, I would never have guessed."

When she led him down the two stairs to the cabin, he was astounded. There were garlands of raisins, cranberries, dried apples and popcorn strung from side to side. And the smell of lobster bait had been replaced by the fragrance of potpourri which she'd been using in their bathrooms. She also had brought several mirrors from home and placed them along the sides of the boat so that they were literally surrounded by glass, expanding the size of the space, yet keeping it cozy. A structural change she'd long wanted: fill the space between the cushioned benches with boards and more cushions making their one bed eight feet wide. Also she'd added another layer of pads and brought silk sheets from her dowry, ones she'd never shown Hal before. She'd also added lights making the cabin brighter than it had been when they first talked there. But she'd changed the bulbs to purple and yellow.

Seeing all of this, Hal exclaimed, "How did you DO all this without my catching you?"

"Oh, once we agreed to marry on the 'Mary Q'...I mean 'Maria D,' dear, I did a little each day, sneaked in the materials when you were doing errands in Statesville, sorta made it a project. Do you like it?"

"It's a miracle and to think it's on a <u>lobster</u> boat, too!"

"Hal, it's my special gift to you, just as renaming the boat is yours to me."

"Does this mean, Maria, that we can't make love on deck in the moonlight?"

"No, dear, I thought of that, too. The long sack there in the corner contains an air mattress we can blow up and take out on deck with a blanket."

"You've thought of everything."

"We'll see. There's one meal on the menu that calls for fish we've not caught. You'll fill that gap! What did you once call it, living with uncertainty?"

Hal laughed, pulled her down onto their new bed and asked, "What else?"

"Now it's your turn to undress me!"

And thus their honeymoon began. They mixed love, fishing, nude swimming, reading, sketching, exploring Jewell Island, luxuriating in the privacy of the lagoon even though another honeymoon boat anchored nearby for a few hours their second afternoon out. Both used plenty of sunscreen, brown though they were from constant exposure to the elements. Only once, when Hal was applying the protective "stuff" to Maria's back, did he hit a sensitive nerve. She winced, exclaiming, "Damn Oliver LaRoux!"

Hal empathized, said, "Sorry." She said no more simply turned quickly and smothered his mouth with kisses.

On Saturday they cruised over to Bailey Island, wandered among the tourists in the gift shop, vetoed the idea of having dinner there, watched a grizzled old lobstermen picking his teeth. Reminded them of Jerry Moulton. He watched them suspiciously, too, then asked rather bluntly, "Newly weds?"

Hal nodded and tried to engage him in conversation, but he looked out toward "The Maria D" to ask, "How many ya got lobsterin' at Baysville?"

"Oh, twenty, thirty."

But he wouldn't talk about Bailey Island fishermen.

Back at their boat, Maria noted, "Sober sort?"

"Yeah, sorta typical, 'fraid uv givin' away trade secrets!"

"Do you feel that way about ours?"

"Trade or love?" Hal quipped.

"Both, silly!"

That evening as they watched the twilight blend into moonlight and lay naked on deck under the moon and stars, Maria asked, "Hal, will you compose a haiku and trace it with your finger on my tummy?"

"That's a lovely request, Maria, let's see..."

And with spontaneity he began writing on her bare skin:

"Orion is falling...

Maria giggled, "Tickles!"

He asked, "What if I don't have enough room?"

"You'll have to improvise."

He finished: "under the moon, into sea
cozy cove sparkles..."

They needed no further foreplay.

Sunday was another sparkling day. They luxuriated with a last, leisurely nude swim. Hal asked Maria if she minded posing under water so he could get some ideas for more sketches. She replied, "I'll pose for you if you'll pose for me."

So for a half hour he swam under, over and around her, in every nameable pose, taking two rolls of film with his underwater camera, only wondering aloud where he could get them developed back in Statesville.

"Better take 'em to Portland. If you feel embarrassed, tell 'em you took the photos for an art class!"

"They'll be terrific shots, Maria, ones to help me understand the human body even more."

"My body, Hal!"

"Of course, dear, but how could I do better anywhere?"

"Better say that, Hal!"

She then modeled "The Thinker," showing him how to pose for that, took pictures of him from every angle before demonstrating how the great sculptor saw "The Kiss." "I'm glad, she said, "there's nobody here to photograph us that way!"

As he held his head back to laugh, she pushed his head under water. He sputtered and spouted as she swam away and he chased her back to the boat where she dived cleanly under the keel to reach the other side. After more hide and seek they climbed up the steps Hal had made for their trip and fell exhausted onto the air mattress, well out of sight. It was an opportunity for more love making.

After brief naps, they went to work to prepare the boat for the trip home...Hal checking the motor, guages and other equipment...Maria giving the gunnels and deck a final swabbing. Their banter was joyful.

Hal noted, "We must return."

"Or come, coming back."

"Maria!" Hal chastised!

"Easy, Hal, keep those ghosts at bay!"

"All right dear! To change the tone: Suppose the fish will miss us?"

"Of course, if we stay on course."

Laughter, gaiety, lightness.

That night at dusk they sailed back into Baysville Harbor, tired but excited about their lives together. Once home, got their gear stored away, they collapsed, noting that after all those hours on the ocean the bed seemed to pitch to and fro as though floating on a ground swell.

Chapter 12

Churning Seas and Stomachs

When Labor Day arrived Maria asked, " Hal, did we hit our 500-trap target?"

"Thanks to you, dear, we came close, maybe 490 or so? We'll never know exactly until we bring some in for the winter and move others out, check 'em off against our list. I've seen dad and Gramps think they were fishing 500 only to discover that they were 25 to 50 short. Also, Maria, it depends upon how many we lose in the fall storms."

"I s'pose," Maria responded, almost absent-mindedly, as she stared off to sea.

Hal sensed something wrong with Maria. The previous two nights she'd gone to bed earlier than usual and resisted his amorous advances. But he kept his counsel, feeling that Maria wanted to keep hers. When he asked, "Is there something wrong? she smiled and shook her head, replied, "Just tired." But he noticed too that she'd not been as free running around the house nude as she'd been earlier in their lives together or during their honeymoon.

But they fished every day as though nothing had happened. Maria held up her end of the task, took turns at the wheel, worked hours at a time with her hands buried in stinky fishbait, washed her clothes daily, "to smell pretty for you, dear." she frequently told Hal. But her defenses collapsed on a Wednesday, the middle of September. She'd done her homework, listened to NOAA radio, reported the weather prediction to Hal, who'd been watching the sky nervously and observed, "I'm 'fraid we're gonna get an early line storm, Maria. We'd better move the traps closest to the shoals at Jocks Rocks or the storm will do it for us. Let's go to bed early and get up early so we'll be out there at dawn. Wind usually doesn't blow hard 'til the tide turns. About ten, did you say, Maria?"

"Yup!"

So they were up at three, doing their chores around the baithouse under the dim wharf light, and rowing out to "The Maria D" about 4. Hal was surprised other Baysville men weren't there. Most had had more experience than he. But once their boat got into the open ocean he understood why. It was already pretty rough.

"Wanta go back, Maria?"

"It's up to you, Hal, you know those ledges better than I. But I'm game to go, dear."

"Tell you what we'll do. We'll haul th' pots closest to the rocks. I'll wheel them in. Keep the coils tight with each trap. We can stack fifteen or twenty on th' stern. Keep your feet clear of the gear in case any fall overboard. It's getting mighty rocky!"

"O.K., dear, tell me what to do, step by step. This is my first in this kind of sea."

"One other thing, Maria. Put on a life preserver. Take my old leather belt out of the tool cabinet, put it through three or four loops in your jeans. Secure with a bowline and take that piece of nylon rope and run it through the belt. I don't want to risk your falling overboard in this churning water. We'd never get you out." By this time the hissing sounds of the waves breaking all around them began to drown out their voices. Despite graying skies combers were white against Jocks Rocks' chocolate brown, and Hal wished his hands were free to write a haiku or two. They had to shout to be heard."

"But, Hal," Maria cried as she finished fixing her lifeline, "Aren't you gonna put on a safety line, too?"

"Can't!" Hal yelled,"gotta be free to move quickly around the deck.   I'll put on a preserver...use my weight to balance against the waves and boat to keep me safe. If I go overboard, plough into the waves as you come up on the port side. The boat won't toss around so much."

"I don't like it," Maria blurted, suddenly dashed for the opposite gunnel, sank to her knees, vomited over the side.

Hal saw her go, stayed the wheel into the wind, rushed to her side, steadying her upper torso. When she regained her composure, he held her arm and yelled, "Seasick?"

She shook her head in the negative, beckoned him back toward the wheel and snuggled beside him, speaking into his ear, "Think I'm pregnant. Missed my period by a week or so."

He looked down at her and held her tightly, "You poor dear, I wish you'd told me this before. No lobster trap's worth risking life in this churning mess. Let's go in."

"No, Hal, we belong out here. We can still rescue a bunch of our traps. Let's stay if the boat can take it." The waves crashing around them were so noisy they communicated with staccato messages.

Speaking directly into her ear and noting that she was beginning to regain some of her pink complexion, he said, "The boat's fine. But nothing must happen to you. I'll tell you a thousand times...I'm happy about the baby...if you're glad. Was it that first time under your bullet-light moon?"

She laughed, nodded, "Time's about right."

"Maria, steady the wheel a moment." He stuck his head in the cabin to get her a thermos of water, held the cup so she could take a swallow or two, embraced her with all his strength, said, "Let's go!"

Within ten minutes, they'd located and pulled five traps. In another twenty, they had seventeen loaded on the deck and gunnels. Hal yelled, "How many more can we take? We're gonna have to take 'em inta harbor. We can't set them in this churning brine."

Maria nodded, held up seven fingers.

Just then Maria happened to look up, saw Digger Danes about a hundred yards away. She pointed and screamed, "Look, Hal, off the starboard. I think he's in trouble!"

Hal looked quickly, yelled back. "You're right! His old engine's been skippin' again. Hell of a time for it to conk out."

By the time they'd reached him, his boat was floating like a dead man in the water, half submerged. He then took a wave right over the side and disappeared.

Hal yelled and pointed, "Maria, cut the rope from that first trap! Leave the buoy on it." She shuffled cautiously about the wet deck keeping her lifeline taut and doing as Hal directed. She tossed the coil to him as she watched for Digger's head.

Hal circled where they thought Digger's boat had been, then to their horror they saw the submerged boat only a few feet off the starboard with Digger hanging to his mast for dear life. His boat was rising and falling with the waves so he would get a breath of air, then go down. Hal quickly gunned his own boat to stay clear of Digger's and not smash his bottom. The wave motion was now so bad that he heard the wheel churn air when the "Maria D" pitch poled over what he estimated must be 10 to 12-foot waves. As he looked back to see if he'd cleared Digger's boat, he suddenly saw Digger pop to the surface, life preserver orange against the black water.

"Maria," he ordered sharply, screaming over the sound of the waves, "Break out the long boat hook. I'm gonna circle back, come up on Digger from the other side. Keep your eye on him. Don't let him float under our wheel."

Hal made a large sweep, keeping his eye on the oncoming waves at one side and Digger's life jacket on the other. He then carefully gauged a run right at him, trying to estimate what the drift of both boat and man might be. He saw fear in Digger's eyes and knew the water was "damned cold." His own face and hands were freezing despite yellow oilskins and gloves. He knew they must depend upon Maria's strength to hold Digger with the boathook until he could stabilize the boat and pull, too. He had one other worry, that Digger's preserver straps might break if a wave struck him broadside as they were pulling him in.

Cautiously, Hal nudged the boat into the on-coming waves, heard the rhythmic churning of his propellor as it surfaced then plunged under water, surfaced and plunged. His engine was laboring but its humming was regular. All gauges were normal, and he shut off all auxiliary equipment to reduce the strain on belts and batteries. All went well for a few moments until they got abreast of Digger. Maria braced herself and held the boathook poised to catch his waist straps, and she could see that his hands were free to grab, too. She hoped he'd not lurch to soon. She screamed above the crashing wave sounds, "Get the hook into your belt, Digger...hook your belt!" But Digger was too eager and jerked away just as she was about to hook him.

Hal saw what happened, left the wheel momentarily to touch her shoulder and reassure her. "Good try, dear, his error. We'll make another pass at him...get him next time." He repeated the circular sweep, aiming to get a little closer, letting the boat drift compensate. And sure enough, Maria hooked his waist before Digger could lurch toward the hook. "Got him!" she screamed and braced herself against the gunnel.

"Hold tight, hold tight!" Hal yelled, leaving the steering wheel again to help pull Digger to the side of the boat. In a few seconds they were hauling him over the rails into the cockpit where he collapsed in a heap and began throwing up. Hal rushed back to the wheel just as a mighty wave crashed over their bow and half filled the cockpit, giving them all an additional shock, the water was so cold.

In the more or less quiet lull before the next wave hit, Hal yelled, "Maria, take the wheel...hold her into the waves."

Maria jumped to it and struggled to hold the wheel steady, Hal dragging Digger into the cabin to peel off his clothes. He was conscious but clearly suffering from hypothermia. Hal tugged against his wet clothing and the rocking boat, cut his pants and shirt off with a fish knife and had him undressed in a few minutes, quickly covering him with a blanket and life preservers.

Rushing back to the wheel, Hal was all praise, speaking directly into her ear, "Beautiful job, Maria. Now we need your nursing skills. Maybe hot coffee and aspirin? Keep him covered up. I'm taking us into harbor." Looking toward the stern, he added, "We've lost some traps, but we saved Digger. We'll jettison all the traps if the boat doesn't respond, Maria."

"The waves and wind are whipping the rudder badly, Hal, hope it'll hold," she ducked through the cabin door to make Digger comfortable.

Although shivering badly, Digger gradually regained some semblance of speech to mutter slowly and thickly, "You thoo thaved my lyth."

"Don't try to talk, Digger. Breathe as deeply as you can. There'll be time to talk later. Here, have another sip of coffee, but don't take it too fast."

Meanwhile, Hal was wrestling with the wheel, clearly realizing that they wouldn't be out of danger until they got well inside the harbor entrance buoy, a black dot bouncing up and down on the waves. He knew it would require a zig-zagging course, taking the waves head on so the boat wouldn't swamp, then turning on a quadrant to take advantage of the power and flow of the waves. At one break in the noise and motion of the boat, he yelled into the cabin. "Turn on the CB, Maria, and call ashore. Tell them to get an ambulance to the dock." Maria did as directed. Digger silently protested by shaking his head, "No!"

Luckily, Maria got Jack Small, told him what had happened, asked him to call the Statesville Fire Department to have an ambulance ready when they got in.

"How long will it be, Maria?"

Maria relayed the question to Hal.

"'Bout an hour or so," Hal yelled back over the sound of crashing waves, checking his watch to see that it was 10:47.

Occasionally they took water over the stern, but the rows of traps, which Hal lashed down during another lull, helped sieve the waves. A couple of times Hal thought they might rip off, move forward and crush him against the cabin wall. When the waves did crash through the traps and sent foamy brine into the scuppers, Hal was glad the boat floated so high in the water. He kept the bilge pumps working in the "Fast" position, silently hoping they wouldn't fail at this crucial point. Fortunately he'd fixed the cabin door at the time of Maria's first coming aboard so knew it was secure against waves dashing directly into the cabin. They'd long since secured the portholes against leakage.

Since giving Maria an arrival time Hal checked his watch every five minutes to determine their progress.

It was 11:30 when they passed through the harbor entrance. Hal was careful to take the "Maria D" right down the middle of the channel. "No point taking any more risks with shortcuts," he reasoned.

When they were clearly inside the harbor, Maria came out of the cabin to stand by his side. "Well done, husband!" she whispered in his ear, kissing his cheek and putting her warm hands into his cold neck. Hal smiled, kept his eyes on the other navigation buoys, put his arm around her shoulders, asked, "How's Digger?"

"I think he'll survive, dear, but he should go to the hospital. He took an awful beating in those waves, and his feet're still cold although I rubbed them."

"And how are you and Junior?" he quipped, leaning over to lick her ear.

"We're fine and will never forget this one!"

"Nor I, dear. You'll recall that dad and Gramps were killed in this kind of weather, the risk of a lobsterman's life...as I warned you."

"And a lobsterwoman, too, Hal?"

"Of course."

"But now all three and a fraction of us are safe?"

A broad smile broke over Hal's face, "Yes, we're all safe!" He hesitated a moment, then leaned over to whisper in Maria's ear, "Do we tell the world our secret?"

She put her forefinger to her lips, then his, shaking her head. "Not yet, I'd better double-check some dates."

Hal nodded knowingly as they passed the last buoy leading them to the dock. There was a large knot of people milling around the ambulance, backed up to the winch on the dock. When the people began waving and cheering, Hal and Maria waved back, Hal using the signal of a fist with thumbs up to indicate that all were safe.

Digger's wife, Annie, and their children were standing at the top of the ladder when Hal eased the "Maria D" against the wharf and Maria secured the lines. Annie shouted down, "Is he O.K., Hal?"

"Got a bad dose of hypothermia, swallowed lots of salt water, Annie, but I think he's going to be O.K. Lost the boat though. Engine failed, and two heavy waves swamped him. Lucky to be alive."

"I told him to get that engine fixed, but to hell with the boat if he's O.K."

Jack Small soon stepped to the edge of the dock, lowered the winch with a stretcher held horizontal on it. He then scrambled down the ladder to help lift Digger out of the cabin and onto the stretcher.

The minute Digger saw his wife, he began to sob. She and the children followed suit. When he tried to speak, Maria hushed him and told him that talking could wait. But Digger wouldn't be put off. In a whisper Maria could barely hear, he said, "Tell her not to scold me."

Maria looked up to the wharf and calmly said, "He's afraid you're going to scold him, Annie."

Annie cried the more, wiping her tears with one of Digger's huge red bandanas. By this time it had begun to rain, a fine mist that clouded the wharf.

Together Hal and Jack rigged the stretcher so it would rise evenly and directed Randy Ballou as he ran the hydraulic winch. The emergency attendants were there waiting when the men swung the stretcher toward the back of the ambulance. Maria was close behind with Digger's clothes, handing them to Annie as Annie gave Maria a hug and a whisper, "I'm so ashamed that I hated you when you first came here."

"It's all right, Annie, we understand. Sorry about the slit clothes and the boat."

"It's all right, Maria, the boat was pretty well insured. Leave it to Digger, he'll be back out there in the next storm. But this time I'm going to force him to get Steve Romney to tune his machine. No need of risking life and limb to save a few bucks with a crotchety old engine." Annie then introduced each of her small children who bashfully shook Maria's hand. She gave each a hug.

Meanwhile, as the ambulance pulled away toward the hospital and Jack loaded Annie and the children into his truck to follow, the lobster clan gathered around Maria and Hal anxiously asking what happened. Hal began, then turned to Maria so she could tell her side of the story. In a few moments Hal spied the two newspaper men from the GAZETTE parking their car and walking rapidly along the wharf.

He turned to Maria to ask, "Should we run?"

Maria saw them coming, too, "No, they did a pretty good story before. Let's not deny our hero and sherohood, Harold Daniels?" He knew she was teasing. So they stood their ground. The crowd parted and made way for the reporter and photographer, Elmer Kruse shouting, "It's a terrific story."

"But, look you guys," Hal said, half kidding, must we tell it over again? It was tough enough to live it, let alone relive it."

Again, Hal and Maria shared the honors and posed for pictures, both on the dock and at the helm of the "Maria D," Hal refusing to let them shoot him alone. One took a quick peek into the cabin where some of the cranberry and raisin garlands still hung. Maria explained, "This was our honeymoon ship."

"You mean, 'Loveboat'?"

"Well, yeah, but not like you're thinkin', the one on TV. Our own private story. But, please, no pictures."

By 1:30 the docks were clear as Hal and Maria moored their boat and began picking up the threads of their lives where they'd left them.

"Much damage to the boat, Hal?"

"Nope, don't think so. But I guess I'll have Steve Romney check the mechanical stuff before we go out again."

"And the traps?"

"These are fine, dear, but who knows about the rest?"

"And your trauma about my morning sickness?"

"You scared me. But what you told me'd better be true."

Maria smiled, added, "When we get home, let's count the days together." They climbed over the gunnels to take the skiff back to the wharf and head for the house.

Chapter 13

Birth and Rebirth

Days turned into weeks; weeks became months. They concluded 'twas true, Maria must have conceived that first night under the artificial moon. They had long heated, even angry discussions about using a medical doctor versus a midwife for delivery, Maria finally achieving a compromise: early medical checks but a competent midwife at labor. She announced it to Baysville neighbors with a bumper sticker for her Omni: HAVE A SPECIAL DELIVERY: PHONE YOUR MIDWIFE. Hal jokingly suggested that she'd better give the hospital a wide berth with that defiant ad. She playfully made a fist, held it under his chin.

At home they began a scrapbook from the newspaper clippings of the LaRoux Incident and their life-saving heroics, also the congratulatory letters they received from prominent members of the county as well as the Mayor of Portland and the Governor of Maine. Hal used the white spaces in the collection to sketch his recollections of various events in their lives. But Maria firmly opposed including nude drawings. "No nudes in that book, Hal. They're private....private, to go with your marvelous underwater shots at Jewell Island. In fact, a separate scrapbook for your eyes and mine alone."

Digger did the unexpected. He and his family had a brass plate made for the Association's meeting room:

FIRST ANNUAL AWARD OF COURAGE
MARIA & HAROLD DANIELS

It was unveiled at the Christmas meeting just before Digger played Santa to present gifts to the children of the town.

Maria occasionally phoned her mother, kept her French alive with such calls. Making notes of the topics as she discussed them, she reviewed them with Hal, point by point, insisting, "I want to withhold nothing from you, dear. I keep telling mom that I can't possibly imagine how our lives could fit together more beautifully. Her return to work's going well. Kathie's doing fine at the University. Mom's cut all communication with Oliver La Roux and his family. She keeps nagging me about church and rearing our child. I keep telling her that the ocean will be <u>his</u> Sunday School and you and I will be <u>his</u> teachers and preachers."

"You're sure, darling, it's a him?"

"Positive."

"Why?"

"Because I feel it, I know!"

At one point Maria asked, "Should we have mom and Kathie for Thanksgiving dinner?"

"If you wish. We can bring lobsters from the boat to save you laboring with a turkey while fishing, too. I'll cook them. You can fix the other stuff."

When Mrs. LaRoux arrived and immediately began speaking French. Maria and Kathie stopped her, Maria observing, "Mom, we can speak your native language over the phone or at your home, but you're good enough at English to use it here. It's only fair to Hal."

Hal chimed in, "I really don't care, Maria, it's O.K. by me. I want her to feel at home here, you can translate what you wish."

"No, dear, it's <u>English</u> here."

Once past that sticky point, the dinner went well. Mrs. LaRoux was very complimentary. The meal got tense only at one point when Oliver LaRoux's trial was mentioned and Maria's mother broke down.

Kathie reached across the table to touch her mother's sleeve, "Not now, mom. Save it for later."

Mrs. LaRoux sniffled, "It's the shame of it all."

Maria got up, walked around the table to give her a hug. Hal looked on compassionately but hardly knew what to say. The conversation turned to the baby, winter fishing and Maria's expecting about May 1st.

Kathie asked, "How long you gonna stay out fishing?"

"Until April 30th!" Maria exclaimed, slyly looking at the circle of faces.

Mrs. LaRoux protested, "*Non! **Non!*** You kill the baby."

Hal looked on amusedly. After much discussion he and Maria agreed that they'd discuss it with the doctor at the end of the seventh month. If she felt O.K., she'd continue as long as possible. Jerry Moulton offered to help every third day for a month or two after delivery until Maria got back on her feet. So Maria explained all of this to her mother and sister, and Kathie stifled a snicker.

When Maria asked about her amusement, Kathie said, "Same old Maria, independent, determined and a kidder."

Family laughter was warm and knowing.

That night after making love they lay in the semi-darkened room. Neither could get to sleep. Finally, with Maria cuddled into the crook of his arm, Hal whispered, "You know, Maria, lots of things puzzle me, but Oliver LaRoux's behavior baffles me most. Frankly, are all Franco men like him?"

"Hal, I can only tell you about the ones I know. Most I grew up with, uncles, cousins and neighbors, they all had chips on their shoulders and would defend themselves at the drop of a hat, especially from anything they thought was Yankee criticism. But they seemed to love their families, their homes, their faith, their language, their culture. No, I think that Oliver LaRoux's an exception, at least among those I knew."

"And is your mother a typical Franco woman?"

Maria remained silent for a long time, so long Hal thought she'd gone to sleep. She stirred, raised herself on one elbow to look into Hal's eyes, commenting, "Am I typical?"

"How would I know?"

"You're sure you had no Franco girl friends before me?"

With that, Hal said, "You're avoiding the question, Maria," and began tickling her. Before long they were both in gales of laughter and holding their stomachs in pain...until Hal did a double-take and pinned Maria to her pillow, whispering, "Can this be good for Junior?"

"I dunno, but what fun," Maria responded, slumping back breathlessly into the crook of Hal's arm.

Meanwhile, daily routine found them rising at dawn, reaching Jocks Rocks about the time their colleagues arrived at their own fishing grounds. As the fall season passed, they moved most of their traps into deeper water, trying to follow the lobsters' migration toward the horizon. One day toward mid-December, Hal looked up from the scrap pads above his wheel, said, "Maria, near's I can tell we lost about twenty traps in the September and October storms."

"Not bad, Hal, we'll build more when we're fishing three days a week. Plenty of time. FARMER'S ALMANAC's predicting a rough season. S'pose we oughta cut the number in half this winter? After all, dear, we're not having to catch so many to break even."

"If you mean, Maria, that we're not gonna set aside your share of the catch as we've been doin', you've gotta another think comin' to you. We've split the housework as evenly as possible, let's split our income the same way. Nobody's gonna exploit you, not even me."

It was a calm if cool day so they could speak in normal tones of voice, so Maria responded, "I love to see people put their money where their mouths are. That's another reason why I love you so much, Hal Daniels. You always touch me with parts of you beyond touch." She left the stern of the boat since they were gently moving between traps, joined him at the wheel, gave him a hug, said, "Look, dear, it's a calm and beautiful day, I'm not too stinky of fishbait...yet. Nice and warm in the cabin. Let's go below and re-enact Jewell Island."

Hal looked into her sparkling eyes, spun around to see where the other Baysville boats were, shut off the motor, pulled the boat into the wind and set the rudder, steered Maria into the cabin. Unbeknown to him, she'd arranged the cushions and blankets to use the cabin's full width. For the first time they made love at work.

Dressed and back on deck, Maria said, "That was our lunch break, dear, no stopping for coffee this morning."

Hal threw his head back and laughed, squinted at the horizon, said, "Better enjoy the memories, Maria, this may be the last calm day until next May!"

"With such delicious mem'ries, I can wait."

The winter passed more quickly than Hal thought it might. The year before he had hated those long cold trips to his father's and Gramps' strings, hated even more handling lines and traps with ice-covered mittens and gloves. And even after Maria came aboard in July, prospects of frigid days haunted him. He dreaded the thought, especially since he felt responsible for Maria, too. Once they were married and she was big with child, he worried even more, expressing such concerns to Maria. She responded matter of factly, "But, Hal, I joined you with my eyes wide open; you said the winters would be tough. Just being here with you is enough. If we have to suffer, we'll suffer together. That makes it bearable."

When he hinted that as a woman, "You shouldn't hafta...," she stopped him, made a fist, stuck it under his chin in a mock punch and reminded him, "No more of that, dear, no more... We're in this together for the duration. I wish it could be forever."

So they halved their fishing traps for the winter, rigging them on strings of ten and tending them every second or third day or when weather permitted. They agreed to make an extra hundred to have in reserve for the next summer. "They'll cost less and we won't feel losses from unexpected storms."

At home they developed a routine which included dreaming, fantasizing their youngster's future, worrying, talking and performing all the body functions to which humans are heir. They both became masters of microwave cooking. Both rushed toward the whistling teakettle when they heard it. Daily, Hal marveled at Maria's delicate touches that turned their weather-beaten old house into a home...fall flowers on their tables, grass heads and heather in tiny vases in subtle places, lily of the valley soap that somehow served to counteract the ever-present smell of fish.

And they enjoyed watching video films rented in Statesville for weekends. At least once or twice a week Hal sketched up a storm, promising to take courses next winter...if they could get babysitters so Maria could also work on her degree. Maria became a master of the circular saw and miter box, making dozens of picture frames, staining or painting them tastefully and filling the house with Hal's sketches. Too, they began picking "favorites" for entering local contests. A Portland gallery sold them like the proverbial hot cakes.

One day between Thanksgiving and Christmas, they were shocked to get a letter from Aunt Sarah, with an "explanation" that she and her husband had been on a trip around the world and had just learned from Sue about their marriage, heroics and Maria's pregnancy. Signing off, she concluded, "I'm enclosing a check for $2000, half for you and half for the baby. I'm damned proud of you, Hal, for carrying on the family tradition. I hope you and Maria will always be happy and have fun spending your part; spend it as you will. Love, Aunt Sarah."

Hal and Maria looked at one another and in unison said, "Let's open a bank account for Junior!" They held the check as though it were unreal.

"And, Hal, with our part, let's start a Jewell Island Fund!"

"Fund or fun?" he grinned.

The winter was cold, indeed, but they managed to fish two or three days a week, got good prices for their lobsters, and grew deeper in bonding and outlook. The spring was typical, slow arriving. Nor did Maria cease fishing when she could hardly climb down the ladder to the skiff and roll over the "Maria D's" gunnels on the newly built steps that Hal rigged for her. They speculated about what Hal should do if she began labor at Jocks Rocks or even farther out. Maria was very emphatic about one other thing: "Boy or girl, Hal, we're gonna bring him out here to fish with us as soon as possible."

Hal shook his head and exclaimed, "You're as headstrong...or as courageous...I don't know which...as my ancestors! Won't we need medical guidance on that decision?"

"Guidance, hell, Hal, we're his parents, aren't we? And mothers know what's best for their kids. I'll breast feed him. We'll love him. You rig a swinging bed in the cabin for him. He'll be fine."

Hal knew it was useless to argue further, simply saying, "I'm glad you're you, dear."

One morning while shaving Hal found himself staring at Maria's toothbrush, hanging beside his own at the bathroom lavatory. It led him to reflect on other staring he'd done since they came together. Just as he couldn't help staring at her nude figure watching the sunrise and tiptoeing toward the bathroom that first morning after, then turning his observations into drawings, he recalled staring at every object she owned or touched as "sacred": her bra, her sweater draped over the cabin door, her fishbait gloves, her socks and underpants tossed on the foot of their bed, her shoes under various chairs. But her toothbrush and comb seemed to "grab" him most. He knew she would let him use them if he asked, yet he felt they were almost beyond touching....something like amulets he'd read about in ancient cultures.

While standing there in a kind of daze, he also spied a single hair she'd dropped on the sink. He was just picking it up when Maria, half dressed, stuck her head through the door, "Hurry it up or aren't we goin' fishin' today?"

Hal gave her a mock salute, told her about his musing.

She responded, "Glad you see all my things as sexy, dear, but they're just 'old things' to me."

"I didn't <u>say</u> 'sexy,' Maria, but that may be the case. I want so frequently to be inside your skin, feel as you feel, think as you think, even toilet as you do, sitting down."

Maria saw that Hal was serious. She tossed her jeans aside, sat on the hopper and drew Hal down into her bare lap putting her arms around Hal's naked torso, studied his face and observed, "Isn't this one of the great puzzles about being alive, Hal? Why can't we be another person, especially if we love them, the air they breathe, the blood in their hearts?"

"A lovely puzzle, Maria, one I feel every day on the boat as I walk through the space you've occupied moments before."

"Hal, my body says, 'Let's go back to bed!' My mind tells me we've gotta go fishing."

"Let's come back to this moment tonight, Maria?"

"It'll not be the same moment, dear, but my tingling skin tells me we'll find the same puzzle if we stare at my toothbrush and bra!"

In seconds they were racing down the stairs.

As she'd promised her family on Thanksgiving Day, she was still baiting traps, juggling them, gauging and rubber-banding lobsters, tossing junk fish to the seagulls and loving every moment of it...on April 30th. But along toward noon, Hal saw her grab the side of the boat, grimace and squat onto her feet. He rushed to her side and felt the fright in her eyes when she looked up at him. She stammered, "Wat-ter broke...Hal...guess we'd bet-ter go in?"

He helped her into the cabin, made her comfortable against the cushions and covered her with a blanket, then went back to the wheel. Through her groanings, he got on the CB, contacted Nettie Small, asked her to call the midwife and get her to the house right away. As he turned to jam the throttle wide open, he heard first one fisherman then another ask, "How's Maria? We're going in with you. Maybe we can help?" From the cabin Maria protested and relayed her wishes that they "keep fishing." But that didn't dissuade them. He looked back to see seven bows pointed right at him. He relayed that news to Maria and saw her begin to cry about their "good fortune."

He made record time getting to the wharf where Jack's wife, Digger and Walt Carmen helped winch Maria to the street level by means of an improvised stretcher.

"How much time do we have, Nettie?"

"First child, probably four to ten hours. She's having contractions every five or six minutes." She turned to Digger, instructing, "Put their boat and lobsters away. Hal, you stay with Maria in the back of your truck. I'll drive you home."

Not long after Hal had undressed Maria and got her settled into their bed, Carrie Barrett, the midwife, came through the front door and ran up the stairs. She took one look at Maria, then turned to Hal to challenge,

"What're you gonna do now?"

"Stay right here, as I told you I would."

"You'll be in the way."

Maria spoke up between groans, "Carrie, please. Hal and I've lived and breathed for this moment. He won't get in the way. He'll help me push," and she grabbed his hand, screwed up her face in contortions.

Carrie slowly calmed, said she expected cooperation, then sent Hal to the kitchen to boil water and fetch towels.

Maria squeezed his hand, nodded and gasped, "I'll be O.K...help Carrie as much as you can. I know you're near."

Maria was in labor for twelve hours, delivering just past midnight, a seven pound baby boy whom they named David Jerry Moulton Daniels. Both Maria and Hal were ecstatic. Within moments of the child's birth, Maria was holding him on her chest and stroking his jet black hair. Hal felt awkward but held his hand over hers barely touching the little one. Carrie took the boy back, bathed him gently, and Maria buried her head in Hal's neck.

For two weeks Hal hardly left her side, waiting on her hand and foot, whether she was nursing the child, helping prepare meals or doing chores, only leaving to buy groceries in Statesville, rent video films or buy seeds and plants which Maria wanted to get into the ground by the end of the month. Finally, she said, "Hal, I'm going to be lonely, but you must get out to our traps. Remember they've gotta be there when I get back to my church, and with the baby."

"I'll go, Maria, but I'll be lonely, too."

"I'll be joining you soon and saluting Jocks Rocks."

As Maria predicted, the weather was cooperative and she was back fishing when David was only a month old. She shocked all the women of the town, but they had come to expect the unexpected from her. They only wondered why they hadn't had the courage to take their own babies fishing. Hal rigged a swinging bed which remained fairly level and steady despite the motion of the boat.. Hence, it was a natural cradle. Thus Maria's heart was filled with joy when she headed back to the open sea. Smiling across the deck to Hal, she sang, she cried to the winds, she let the breezes and sun sweep through her newly-cropped hair; she held up both arms to the seagulls circling "The Maria D,"   warning, "Don't dump on our baby!"

Hal, too, was joyful, noting rhetorically, "Love being back in church?" She did a tiny shuffle that he interpreted as a dance. "Overjoyed, dear!"

The summer was like that. Day after day they carried David, now "Dave" but sometimes "Jerry," to the dock at dawn, adjusted his schedule to their fishing routine. She fed him upon demand, nursing him both on the open deck and in the cabin, depending upon the weather. And he grew like the proverbial weed. Fishing went well, too. Their fishing friends frequently pulled along side at Jocks Rocks and asked to see Dave. Women in Baysville sometimes "slipped down to the docks" when they saw "The Maria D" round the bend and sail into harbor. In fact he became a kind of symbol of good feeling and luck in the community. Some called it a "sea change." Daily news of David and his parents traveled rapidly to Barnsville; Jerry Moulton saw to that, he was so proud of his "namesake." Even some of the Barnsville gang flew a special flag Jerry designed (so there could be no miscommunication) and sailed their boats to the Baysville fishing grounds to see this baby who had become an almost instant legend. The Statesville GAZETTE did a story on "The Fishing Baby," which the proud parents added to their scrapbook. They kept a joint baby journal, too, one that Hal illustrated from time to time, usually catching David's chubbiness and happy smile. Hal and Maria only worried that the child would become so spoiled they'd be unable to reach or teach him when he got older. Hal's and Maria's bonding became ever stronger as they nurtured David and one another.

They welcomed occasional visits by Maria's mother and sister and frequently slipped Kathie money for books and other personal needs as she worked her way through some of the tougher moments of her pre-vet courses at the University. And they survived Oliver LaRoux's trial, appearing only briefly when LaRoux's lawyer convinced him to plead guilty to shorten his term. They felt relieved when he was sentenced to "ten years with no parole" in the state prison, but quietly speculated that AIDS would "get him first."

That winter, Maria consented to leaving David with Nettie Small who was attempting to begin a day care center. Maria was emphatic, "But only one day a week, Hal. I can't leave him for more than that. You fish one day alone and one with me. If need be, we can always fall back on our savings, just for this one winter. Next winter..."

He repeated, "Next winter?"

"I dunno, we'll see what happens?"

The following summer they spent a couple of months studying school catalogues and debating what action they should take. They applied and were admitted to Portland Art School and University of Southern Maine. They developed a routine of being back at the docks by noon on Mondays, leaving David on alternate Monday afternoons with Maria's mother or Priscilla Moulton, his namesake's wife, on their way to Portland. This led to pressure, some petty bickering and their first major argument, not to mention a near fatality on I-95.

Neither liked to "impose" on their families to care for David. They both got overtired and finally agreed not to fish on days after school during the Winter Semester, no matter the weather. But they amazed themselves at the intensity of their arguments, usually in the Omni returning home after classes. Disagreement was over the processes and purposes of attending classes.

Since both wanted to express their feelings the minute they started home, some of their frustrations grew from who would be first, who was listening, and what to do about their discomforts. After a couple of heated discussions they agreed that Maria would get "first crack" at describing what had happened to her. It began the third week of their evening seminar schedule.

Maria had been given junior year status and decided to major in Education and the Social Sciences, so she signed up for a Modern American history course and Introduction to Political Science. On a cold night late in January she got into the car, boiling with anger. Hal could feel it coming. She began, "Of all the gawdamn bullshit that jerk Jones slings. He sticks to the textbook. Never

recognizes any of our hands. He claims to be objective,
isn't. He gives us the standard Republican Party line. Hal,
I don't know what th' hell I'm doing in that course,
degree or no degree. I'm not sure it's worth all the
sacrifices we're making in time and money, coming in
here to hear this shit!"

"But, Maria, didn't we agree to reassess the
situation at the end of the term? No reason why we can't
do that."

"O.K., O.K., I'm going after a degree to give us
security if anything happens to you. But, Hal," she
screamed in the close confines of the car, "it may drive me
nuts!"

In a few miles she'd calmed down enough to ask,
"And how did your classes go?"

"Mediocre. I'm getting theory and history I didn't
know, but the practical stuff is pretty elementary. I forgot
all that guy Leary knows when I was in the third grade."

"So you, too, have reservations?"

"Of course, dear, but none of this should mean the
end of the world. We have our own world, and it's a
beautiful one." With that he put his hand over hers, then
massaged the back of her neck as she drove.

"Is that foreplay, Hal?"

"Not really, dear, just telling you I love you
through my finger tips."

The next week, as Maria started to climb in to drive,

she sloshed through a snowbank at the edge of the school parking lot, went over her boots, cursed, went around to the passenger door to ask, "Hal, do you mind driving?"

As Hal slid under the wheel, he could see that Maria had been crying. But he started up the Omni, which he didn't particularly like for its "oversteering," and headed for Baysville. They hadn't gotten free of Portland's slippery streets when Maria began taking her history professor apart. But this time Hal could see that she was dealing more with his attitude than the content of the course. Toward the end of her diatribe, she stopped short after saying "and...."

"And...?" Hal questioned.

Maria released a flood of tears.

Hal comforted her with his free hand as best he could until she stopped sobbing. She looked up, her eyes shining in the oncoming headlights, continuing, "And that son of a bitch had the colossal nerve to make a pass at me when I went to his office to ask him to clarify a n obscure passage in the essay on method."

"He did!" Hal exclaimed, pulling over against a snowbank at the side of the road. "I'm tempted to go back and punch out the bastard!"

Maria was now confused, but went on in short gasps. "Oh, please don't, Hal. It's too late in the term to stop now. You'll get into trouble if you hit him. One of my classmates tells me that these profs, you know, are untouchable, especially if they've got academic tenure."

"What's that, Maria?"

"Can't be fired. Drive on home, Hal, I'll get over it."

"But I may not!" Hal started down the road to the Turnpike, continuing the conversation. "D'ya know anything about this guy's reputation? Does he do that to all the women in the class?"

"I don't know, dear. I don't have time to talk with many of them. Maybe he misread my visit. He may have thought that I was making a pass at <u>him</u>, my question may have seemed so simple? I'm confused. Let's stop talking about me and talk about your class."

"Routine. The guy pisses me off. He's good on history and theory but doesn't know beans about drawing. He says almost nothing about the assignments I do at home, hasn't given me a single suggestion, good or bad, about my sketches. You know how I work on them ev'ry week. I dunno either."

Maria tried to comfort him, and they talked about their classwork as part of what they called "the long haul."

But not long after reaching the Turnpike a semi passed them at high speed, the wind and snow hitting the Omni broadside to make it fishtail. Hal wrestled with the steering wheel as Maria screamed. They slammed into the hard snowbank the ploughs had made beside the road, hit it going twenty-five or thirty miles an hour. Within what seemed like seconds a State Police car pulled up behind them.

The door was jammed on her side of the car so she slid out under the steering wheel. It was close to ten degrees with a light snow blowing. The policeman asked to see Hal's license. No problem. Suspecting Hal had been drinking, he directed him to walk a few steps, which he did. He finally decided to check the damage of the car and ask what happened. The officer was polite, wrote it up, asked what Hal and Maria wanted to do. Hal said that they would continue if the car ran properly after he got a shovel and dug it out of the bank. In a half hour or so, Hal freed the car so Maria could drive it forward as he and the officer rocked it out of the ruts.

While the car was badly bent up, Hal slid under as far as he could, flashed a light to check the front end, reporting, "I think we can probably crawl home, officer, if we can get off onto side roads. We just came from the University and need to go to Barnsville to pick up our child. Would you be willing to use your CB to tell the Jerry Moulton family that we've been in an accident, are O.K. and will be an hour or so late?"

The officer obliged, driving behind them with a flashing blue light until they could exit the Turnpike and drive to Barnsville via back roads. During the drive Maria was contrite, blaming herself for upsetting Hal.

Hal would have none of it, pointing out that there was no connection between the speeding semi and their states of mind following classes, concluding, "No, Maria, neither of us need this as an excuse for dropping our classes. We're almost half way through them. Let's finish, then decide what to do."

Maria replied, "Hal, wonder what our crooked road will tell us tonight?"

"Maybe to stay home and go to bed on cold snowy nights!" Hal quipped, pulling her closer to him.

Both completed their courses, using the pick-up to travel back and forth to their classes. They also decided to trade in the Omni, without fixing it, as soon as spring arrived. Maria gave her history professor a wide berth, studied hard, got low B's on her exams and no further harassment. Hal did a perfunctory job in his classes, made C's on his exams but A's on his drawings, also getting a B for his course.

The most intense argument in their lives together grew from discussion over whether or not to continue. Hal seriously probed the matter of the college degree, said that Maria was too good a lobsterwoman, as both he and the boys at the dock agreed, to take time from being a mother to go to classes to listen to "assholes" pontificate about history and politics. "Don't we agree that American politicians are do-nothings?"

All these arguments sent Maria sailing into the east wind! "Of course I'm a good lobsterwoman, dear, but what are you doing talking with the boys about my education? If you went the way of your father and grandfather, would those guys help me and David?"

"They love you, Maria. Of course they'd help you."

"I'm not so sure how deep, Hal, how deep?"

"Maria, I'm not planning to commit suicide or go away. These guys have turned out to be decent human beings. And there's no place here in Baysville for a college-educated woman. Anyway, you'd have to commute to Statesville or Portland if you wanted to use your degree, dear."

The disagreement sometimes got so intense that Maria wouldn't speak to Hal for days, at home or in the confines of the "Maria D" until one night after passionate love-making, Maria looked up into Hal's weather-beaten face, remarked, "O.K., I've been stupid and stubborn. I don't feel good about our silences. Let's declare a truce. No more talk about college or classes until David grows up. If need be, maybe I can sign up for one of those university without walls programs. Now, it's you, David and fishing. What say?"

Hal's eyes welled up as he buried his face in Maria's neck.

And thus the family fished for living, love and happiness. David grew into a toddler, spent warm days on deck, "helping" his mom and dad, usually at the end of a tether with a special harness Maria sewed for him. Among his first words were "bowoot," "seagal," "wope," "lobshter." He loved to hear the lines whining in the open pulley above the winch and invented his own sound simulating it. Sometimes he came close to disaster as he sat on the dashboard above the wheel, dangling his feet dangerously close to the hydraulic mechanism....or stuck his fingers toward a lobster claw before Hal or Maria could secure it with a plug or rubber band. But they watched him like proverbial hawks, and Maria frequently hugged him with the comment, "I could love you to death." Only later did this comment come home to haunt them.

It was May lst, David's second birthday. Maria was in the kitchen making a cake after sending Hal to the store to buy ice cream. For just a moment she neglected the boy. He heard his father's truck coming down the road, slipped quickly through the screen door, ran to hide in the bushes as he'd done before with Maria at his side, "Surprise daddy....surprise daddy." This time he ran out just as Hal pulled into the yard. The truck ran over him and killed him instantly. Hal was out of the truck in a flash. Maria flew out of the house upon hearing the impact. When they saw him lying behind the truck, they picked him up, both wailing as though the cosmos itself had cracked.

They saw that he was dead and slowly carried him to the house, placed him on their bed, and wept for what Hal later recalled as "hours."

Over and over Maria cried, "Oh, Hal, why did I lose track of him? Why? Why? Why?"

Hal blamed himself for not seeing "the little tike" hiding in the bushes.

Each consoled the other, over and over....even before calling the authorities. When the townsfolk learned of the tragedy, they flocked in to hold both parents in their arms. And even after the simple funeral, held in the Fishermen's Association meeting rooms and bulging with the entire community, they poured into Hal's and Maria's home with food enough to feed an army.

Despite Maria's mother's clamoring for a traditional burial, they had David's body cremated and asked Jack and Nettie Small if they would hold the urn until August lst when they wanted a Memorial Service near Jocks Rocks...on the very anniversary of his being conceived. Jack put his arm around both their shoulders, saying, "Shouldn't you do it sooner? Three months is a long time to put it off when you should begin to forget?"

Maria spoke up, "Thank you, Jack, you're a dear to think of us like that, but we'll probably never ever get over it. I'm going back to fishing with Hal tomorrow, but it will be comforting to know that he's in your good hands, just down the street. Please take Dave's urn home with you."

Jack obliged, wiped tears away from his own eyes, gave her a hug, kissed her forehead.

The next three months were, indeed, difficult. Hal and Maria experienced many moments of silence; each knew the other was thinking of David. But they smiled at the world, and the people of Baysville and Barnsville smiled back, amazed that Hal and Maria could bounce back to fish, to laugh, to attend meetings, to help others in need.   And while Hal's and Maria's faces frequently showed their extended bouts of crying in one another's arms, they continued their loving embraces, too. Whenever Maria was discouraged or depressed on the dark of the moon, upon undressing for bed, she would put on Hal's bullet lamp and joke about "your full moon."   And on full moon nights, she would wander nude around their bedroom observing that their cottage was a kind of moon dial, telling them the time of night as the window frames cast dark shadows along the floor. Hal would lie in a state of excitement, waiting for her to come to bed, then suddenly jerk to a state of wakefulness, knowing he had been mesmerized by staring at her beautiful body as she literally danced from one window to another.

It was the Sunday before the Fourth. They luxuriated in bed with the knowledge they didn't have to go fishing. Suddenly Maria sat upright, her breasts firmly pointed toward Hal, and said, "Hal, I've got it. We'll put

David's dust in 'The Maria D,' sail out to the far edge of Jocks Rocks, hold a memorial ceremony for him out there. You arrange the music and think of words we can say together. We'll go alone. We know and feel our tears. Can we pull one trap in his honor, no more than one that day, no more...?" Then she fell onto his bare chest, and sobbed body-wrenching sobs. That afternoon, for the first time since David's death, they walked along the beach, hand-in-hand, occasionally picking up a dead crab or skipping a smooth stone along the surf, counting the bounces in unison. Otherwise they hardly spoke but knew the other's thoughts, felt the other's feelings. When they got back to the house, Maria insisted that they go to bed and make love. Hours later when they awoke to discover the sun, a red ball, just dipping through the oak trees across the marsh and yellow light turning the sea spray into golden fountains, Maria said, "Hal, I've had my catharsis. No more until David's memorial."

Hal said nothing, just pulled her close to him and hugged her tightly 'til they went back to sleep.

On that August 1st morning Hal and Maria stopped at Jack's and Nettie's house about eight. They'd phoned Jack the night before. Jack asked if they'd like for him to come along with the urn. They declined politely, told him they had to go by themselves, thanked the Smalls "for everything," gave each a round of hugs.

Maria carried the ashes in her small satchel, and they drove to the dock. Following their routine for getting under way, they hit the open sea about nine o'clock. It was calm and sunny, but they saw their colleagues breaking out their tiny stabilizing sails so knew it was starting to blow outside. Both were dry-eyed. They'd agreed to remain silent. "Words would get in the way." Hal prepared the music, and Maria asked him to dump David's ashes on the water as she skippered the boat.

A half hour later Maria took the urn from her satchel, lifted it to her lips, then handed it to Hal as she replaced him at the wheel and began circling their closest trap. Hal snatched the buoy with his long boathook and handed her the rope. She threaded it through the open pulley of the winch, was sensitive to the whining of the equipment and quickly brought the trap to the surface. Both stood quietly with heads bowed for a moment, not opening the trap though they could see it filled with life. Silently, she nodded to Hal and together they shoved the trap into the swirling waters. He sprinkled the ashes into the water behind the trap as they both watched some of the ashes swirl into the vortex while other bits floated under the boat toward Jocks Rocks as a lone seagull kited overhead, uttering its most poignant mewing cry...like some lost soul far out on the edges of the cosmos. Hal reached over to start the tape recorder as they put their cheeks together, their tears flowing as one river.

In a few moments both returned to the wheel, Maria's hand held over Hal's. Half way to shore he asked Maria to take the wheel while he went into the cabin. She thought he was going to the head; but instead he rapidly scribbled on an old yellow pad, tucked the paper into his breast pocket, and returned to the wheel to find they were nearly back to the harbor. Holding Maria close and occasionally rubbing his head against hers, he took the "Maria D" back to the mooring as they reversed the day's routine.

Almost back to the house, she said, "You wrote something, didn't you, dear?"

"How'd you know?"

"Saw it in your pocket. A poem?"

"Yes, want to hear it?"

"Not now, dear. Let's go over to our favorite crooked road? I'll drive so you can read it to me. I've got something to tell you."

"Of course, we'll switch driving at the public beach."

Ten minutes later Maria was at the wheel and Hal pulled the rumpled paper from his pocket and slowly read his poem just as it had come out of his feelings and pen:

>As the fog burned off
>over our lighthouse
>we reached the dock
>greeted old ghosts
>with familiar
>"Hi thayuh's" and "Ayuhs"
>discussed windy weather...
>
>tossed gear aboard
>noted tide flow past the spindles
>took "The Maria D" out to sea
>as the sun popped bright yellow
>over dim cottage cubes...
>
>hit Number Two Buoy
>by the time engines smoothed into
>regular exhaust rhythms and
>decks began to dry....

sliced 'tween Little Gull Isle
and inner Jocks Rocks
setting due southeast compass
noting lobster buoys tugging shoreward
in the flowing tide...

both silent
feeling more than thinking:

"Such a big vessel
for carrying so small a package"

        ("Ashes to ashes and
                    dust to dust..."
...was all I could muster)
sun glist'ning on the low groundswells
of water and emotion
pale lemon-yellow light eerie
gull following/wheeling
as though on death patrol...

'bout a mile out, homes and lighthouse
still sporting wisps of fog,
like scarves, white wreaths,
for this little boy
who darted under his father's truck...
tears welling in mother's eyes,
my own burning as we brought the vessel
to our agreed-upon destination
facing the flowing waves, to keep
our promise to David and ourselves...

boat slowing
gull circling, mewing her poignant mew

      pull trap, set it back in swirling brine
      start tape recorder and sprinkle
      THE TINY HANDFUL OF ASHES
      on the whirling waters
      tinny organ music
      weakly counterpoint
      to pounding waves
      on a nearby reef
      our grief expressed in

            "Nearer My God to Thee"
                      &
                "Finlandia"

      boat lurches, stomach churning
      as we come about to head home,
      silently looking back
      our half-masted flag whipping
      noisily over the deck...
      honor to a bright little boy
      who would not grow up
      to steer
      his father's and mother's lobster boat...and
      test the beauties and dangers of the sea....

When Hal finished reading, they were both crying, and Maria abruptly announced, "I've gotta stop, Hal, I can't see." He reached over to steady the wheel as she pulled into a dirt road beside the crooked highway.

"It's a beautiful creation," she cried, between spasms of tears. He pulled her head to his shoulder.

After a long silence and the passing of many cars, she raised her head from his shoulder, took his chin in her hand, looked into his eyes and blurted, "Hal, I'm pregnant."

Hal smiled through his tears, "Boy or girl?"

"Does it matter?"

"Not really!"